SPACE WEAPONS/SPACE WAR

SPACE WEAPONS
SPACE WAR

John W. Macvey

STEIN AND DAY/*Publishers*/New York

First published in 1979
Copyright © 1979 by John W. Macvey
All rights reserved.
Designed by L. Hebard
Printed in the United States of America
Stein and Day/*Publishers*/Scarborough House
Briarcliff Manor, N.Y. 10510

Library of Congress Cataloging in Publication Data

Macvey, John W
  Space weapons/space war.

  Bibliography: p.
  Includes index.
  I. Title.
PZ4.M1782Sp [PS3563.A3356]   813'.5'4   78-24147
ISBN 0-8128-2579-9

Figures 5A, 5B, 6, 6A, and 6B are reprinted from the author's *Interstellar Travel,*
published by Stein and Day.

To My Parents

"The hour is very late, and no one can guess how many strange eyes and minds are already turned upon the planet Earth."

ARTHUR C. CLARKE

CONTENTS

CONTENTS

ACKNOWLEDGMENTS

The author wishes to acknowledge with considerable thanks the services of all those who helped to make this book possible, and in particular two young ladies: Mrs. Isobel McCaa, who quickly and efficiently typed the script, and Miss Helen Campbell, who once again produced the various diagrams and illustrations.

Recognition and thanks are also due in full measure to the editor, editorial staff and management of Stein and Day who, at all times, and in many diverse ways, guided the manuscript in its passage to final book form.

Saltcoats, Scotland   JOHN W. MACVEY
1979

PROLOGUE

The main activity of the highest living organisms in the Universe can be also the colonization of other worlds.
Konstantin Eduardovich Tsiolkovsky, 1934

There is a good chance that, among the, say 100 stars closest to the Sun, some have planets bearing life well advanced in evolutions. The chances are then good that in some of these planets animals exist evolved much farther than men. A civilization only a few hundred years more advanced than ours would have technical possibilities by far greater than those available now to us.
Giuseppe Cocconi, 1959

One in one hundred thousand stars have advanced societies in orbit around them.
Carl Sagan, 1961

We must expect most galactic societies that have crossed the threshold of civilization to be far more advanced than our own.
Frank D. Drake, 1960

No one would have believed ... that this world was being watched keenly and closely by intelligences greater than man's and yet as mortal as his own. Yet across the gulf of space minds that are to our minds as ours are to those of the beasts that

perish, intellects vast and cool and unsympathetic, regarded this
Earth with envious eyes and slowly and surely drew their plans
against us.

H. G. Wells, 1898

I know perfectly well that at this moment the whole Universe is
listening to us—that every word we say echoes to the remotest
star.

Jean Giraudoux, 1945

The vast distances between solar systems may be a form of
divine quarantine.

C. S. Lewis, circa 1950

Even radio contact with a superior civilization would lead to
profound upheavals.

NASA Committee on Long Range Studies, 1960

Would beings from another world covet our gold or other rare
substances? Do they want us as cattle or as slaves?

Ronald N. Bracewell, 1962

One of the primary motivations for the exploration of the New
World was to convert the inhabitants to Christianity—peace-
fully, if possible; forcefully if necessary. Can we exclude the
possibility of an extraterrestrial evangelism? ———Can we conclude even darker motives? Might an extraterrestrial society
want to be alone at the summit of galactic power and make a
careful effort to crush prospective contenders? Or might there
even be the 'cockroach response'—to stamp out an alien creature
simply because it is different.

Carl Sagan, 1966

Intelligence may indeed be a benign influence creating isolated
groups of philosopher-kings far apart in the heavens and enabling them to share at leisure their accumulated wisdom. On the
other hand, intelligence may be a cancer of purposeless technological exploitation, sweeping across a galaxy as irresistibly as

it has swept across our own planet. Assuming interstellar travel at moderate speeds, the technological cancer could spread over the whole galaxy in a few million years, a time very short compared with the life of a planet————. It is just as unscientific to impute to remote intelligences, wisdom and serenity as it is to impute to them irrational and murderous impulses. We must be prepared for either possibility.

Freeman J. Dyson, 1964

In a direct confrontation with superior creatures from another world, the reins would be torn from our hands.

Carl Gustav Jung, early 20th century

PREFACE

The theme of weapons and space war may seem a rather odd and unhappy one in the closing decades of a century that has known more terrible and destructive wars than any other. These conflicts have ranged from the hideous paroxysms of two world wars through a whole series of ever more sanguinary minor wars. "Minor" has unfortunately become a very relative term now that the lethal power of weapons has so increased.

We might hope therefore to be done with war, to have learned that in the end violence solves nothing and almost certainly creates newer and greater problems. This should not be taken to mean that a nation wantonly attacked by another should not defend itself to the best of its ability and resources. There will always be much to be said for dying a free person rather than living as a slave. In view of some of the appalling political ideologies rampant during the century, this outlook must be commended. Freedom is a precious thing. Nevertheless, our world, apart from a few demagogues and political adventurers, is a world heartily sick of war—one longing for an enduring peace, and never has this been more essential, for humanity as a whole now faces immense and growing problems. The resources of our planet, once seemingly abundant, are realized now to be all too finite. Its minerals, its metals, its fossil fuels are non-renewable resources. Once they have been exhausted there are no more, and no ostrich head-in-the-sand attitude or political platitudes and double talk can possibly alter that. (In the strictest sense these things *are* renewable, but only on the geological time scale where a million years is as a day.) At the same time the population of our planet continues to grow by leaps and bounds and each person, not unnaturally, sees no reason why he or she should not have a rightful share of the world's gifts. It is a pattern for disaster of an unprecedented kind. The problem becomes basically one of simple mathematics. If the planet's resources are to be shared by x persons, there may be enough for all for a protracted period. But if they are to be divided among 2x persons, the amount available for each is effectively halved. And if 2x becomes 3x, or 4x, or more, the problem becomes one of alarming proportions. War may trim the world's population, but war also consumes in a very avid way its increasingly limited resources. In time population loss is made good, but nothing can atone for the material wastage.

Why then a book on the theme of war? Today man has already put a first hesitant, faltering footstep into space. It is the first step on a road that, God willing, may in time take him to the stars themselves. This has long been the dream of mankind, epitomized magnificently in the stirring lines spoken by Cabal in the closing scene of the film version of H. G. Wells's classic and memorable *Things to Come*.

> For man no rest and no ending. He must go on. Conquest beyond conquest. First this little planet with all its winds and waves and then all the laws of mind and matter that restrain him. Then the planets about him and, at last, out across immensity to the stars! And when he has conquered all the deeps of space and all the mysteries of time, still he will be beginning. . . . It is this, or that. All the universe or nothing. Which shall it be?

With such words ringing in our ears it is easy to carry this great and noble philosophy even farther—to perceive of our linking up and learning from other, older and infinitely more mature cosmic communities within our galaxy. And, of course, eventually this might well be. The future for the terrestrial race out there could be bright beyond belief. Homo sapiens, born and nurtured on a lowly planet toward the rim of the galaxy, might yet become galactic man with all the status and wisdom and power that the title implies. His civilization could be merged with the greater civilizations of others so bringing about the long-sought "millennium." It is a wonderful dream; it is also one that could turn into a nightmare. This is a possibility that must be faced. We cannot, if we are going to be realistic, refuse to concede certain eventualities. Here on Earth, for countless centuries, man has fought man; tribe has fought tribe; nation has fought nation. In space we must allow for a progression of this sequence, in which planet fights planet and star fights star—however unwelcome and repugnant that might be. Evil and unpleasant things do not go away merely by a refusal to look at them. The fact that we recognize such possibilities does not imply either endorsement of them or a pessimistic belief in their certainty. On the contrary, we hope fervently that such events will never come to pass—but we cannot, in truth, say they could *never* be.

The pressures that in time could force us to seek a new world for our kind among the stars could be the same pressures already compelling some of our neighbors in the galaxy to do just that very thing. These pressures could, in the end, be of a kind too strong to be restrained by reason of ethics. There is no point in being foolishly idealistic. These things could be—and in our hearts we know it.

The pages that follow neither support the concept of waging interstellar war nor proclaim the writer's belief in the certainty or imminence of its coming. They *do,* however, support his belief in its *possibility* and the desirability of considering its consequences and their awful implications.

Our galaxy, in common with all other galaxies, is unimaginably vast. Or perhaps this is just an illusion arising from the fact that we are so small. Either way the relative proportions remain. But the vastness should be seen neither as a barrier to our exploration of the cosmos nor as a shield behind which we can forever crouch in absolute safety and exclusion. Out there in the star mists could be other races. Out there, too, could be forces and fleets quite beyond our imaginings. Our world, despite all the abuses we have given it through the centuries, is still a very good world. It is also *the only one we have got.* We must defend it if we are ever attacked from space, for the very good and starkly simple reason that we have, as yet, nowhere else to go.

We hope such a day will never dawn. That it will not can never be a certainty.

John M. Macvey

1. BUT WHY WAR?

In the realm of science fiction the theme of warfare has, from the beginning, been a continually recurring one. For the most part it has tended toward *space* war, generally between terrestrials and beings from other planets. As science fiction developed and grew more popular the canvas was extensively broadened by the concept of interstellar and intergalactic warfare. So was born the world of Flash Gordon, Buck Rogers, and other stalwarts, male and female, of the space ways, though it must be admitted that fiction here showed an increasing dominance over science. But then, who cared? It was all great rousing, exciting, and entertaining stuff. In the depths of space great fleets battled it out with colored death rays, disintegrating beams, and other "refinements" of supertechnology.

In my youth and for my generation this was all relatively new, so that pulp science fiction with the most lurid of covers became a way of life—and a source of despair to schoolteachers of the time doing their best to inculcate in us the merits of English literature. Each episode of Flash Gordon's adventures at the local movie theater (generally screened on a Saturday afternoon to a highly expectant and vociferous audience) was received in an atmosphere of highly charged excitement and emotion. Would Flash save the lovely Dale from the clutches of the wicked Emperor Ming? (He always did, but by the close of the next episode the poor girl was generally in even more dire straits.) Being young and then very impressionable, much of this rubbed off. I clearly recall a period around the age of fourteen

1

when several of my contemporaries and I decided that we, too, would create science fiction epics. With the boundless energy and enthusiasm of youth we quite firmly believed that under our respective pens sci-fi classics would rapidly come into being. Alas for all the dreams! Most of the "epics" never saw completion; the few that did were laid aside and eventually forgotten. But great was the pleasure and excitement derived in the creation of those figments of our boyish imaginations. For weeks on end we lived in an ecstatic dream world of other planets, weird spaceships, death rays, alien peoples, stalwart heroes, and lovely heroines. And in this world we ruled. The heroes, the heroines, the lot were at our bidding. They were the puppets and we held the strings. I fear that round about this time scholastic efforts tended to suffer; Euclid's theorems and French irregular verbs had little chance against such heady stuff.

I can still remember, almost as if it were yesterday, my own particular saga which, with great unoriginality, involved the first space trip to Mars. Our near neighbor, the Moon, was ruled out as a destination. It was dead and lifeless—hardly a domain for an exciting space opera. True, a fellow called H. G. Wells had done all right on the Moon. But he had changed the Moon to suit his own purposes, even going so far as to produce lunar beings called Selenites. This, I felt, wasn't quite right. At that time (the late 1930s) Mars was still the undoubted abode of a highly advanced and scientific race of beings. It had a vast network of canals and other futuristic marvels. Here then was the only possible setting for an adventurous space epic. The Moon was a mere pock-marked cinder, which was disposed of in half a page as the occupants of my spaceship, the *Gloria Mundi* (I feel the name owed not a little to my Latin master of the time), surveyed its unpromising and stark surface on a close flyby. I cannot remember too much about the characters now, save that they numbered six, one being a girl named Julia, who just had, of course, to be the leader's attractive daughter. At the tender age of fourteen I was a bit undecided what ultimately to do about this young woman. Presumably I could marry her off to another crew member. And in between she could be rescued from all sorts of quite ghastly situations. Naturally Mars received the full treatment; canals abounded, the atmosphere was breathable to terrestrials, and the lower gravity gave a little trouble—but only at first. Needless to say there were Martians, not at all grotesque (at least I spared my readers the "bug-eyed monster") and they, of course, had a very highly advanced civilization. Inevitably, too, they had ray guns—red, yellow and blue beams. I had a strange predilection for the primary colors. And the ray guns could, among other things, quickly transform rocks and boulders into steaming, sulfurous pools of bubbling magma. In some respects, oddly enough, I did not leap too far ahead technologically. Martian aircraft were simply superhelicopters (of course, in 1937 such craft were virtually unknown) while short-haul intercity transit was by monorail cars on elevated tracks—true monorailers riding a single flanged track and balanced by internal gyroscopes—none of the present guided aircushion stuff. Part of my Mars was ruled by the kindly and benevolent Emperor Hwango of the state of Unxora. The aim in selecting personal and place names was to go for something as bizarre and strange-sounding as possible. In this at least I was successful. Inevitably, he was in dispute with another of the Martian superpowers, which was ruled by a corrupt and ruthless tyrant. Hwango, predictably, had the help and cooperation of his terrestrial visitors. At first the language problem worried me. Eventually, however, I found the answer that so many other (and more successful) science fiction authors have found since: the Martians had become conversant with the languages of Earth by the simple expedient of listening to terrestrial radio broadcasts. Why such technologically advanced beings had not already contacted Earth by radio or created spaceships of their own to go there did not trouble me. And so the scene was set for internecine Martian warfare, which swept over the surface of the planet and out into space itself. Naturally in the end good prevailed over evil and everybody lived happily ever after with the exception of the tyrant and his henchmen, who got their just and long-overdue deserts. It was all good space opera—or perhaps more correctly—a western set on Mars (the Seventh Cavalry rode helicopters but they arrived in the nick of time just the same). My ideas and visions at that time, though perhaps prompted by the earlier writings of Wells, were much more influenced by the famous "Martian" novels of the prolific American author Edgar Rice Burroughs.

The idea of interplanetary war and its true terrors was handled magnificently by H. G. Wells in the closing years of the nineteenth century in one of the earliest and still probably one of the best of science fiction classics, the now almost legendary *War of the Worlds*. Here is no mere space opera, but a serious attempt to describe the horrors of an invasion of our planet by intelligent beings from another world. Wells chose Mars presumably because it was close, relatively speaking, and also, of course, because at that time Mars was seriously regarded, even in some scientific circles, as a likely abode for a highly advanced alien society. That this was so is due in no small measure to a number of astronomers of considerable repute—Schiaparelli, Lowell, and Proctor, to name but three. The very word "Mars" had a certain ring to it; it meant Martians, beings of vast intellect on another world. Today it merely means the small desert-surfaced fourth planet of the sun. Rather a pity, I suppose.

To a very large extent this book bears the imprint of its own time; yet somehow this augments rather than lessens its impact. Terrestrial society was then much more leisurely. Transport was dominated not by the internal combustion engine but by the horse and the steam railroad engine. Communications were still slow. Radio was unknown and Wells speaks freely of the electric telegraph and even the heliograph. So in that quieter, slower age the contrast with the ruthless, technological Martians is much more marked.

In Wells, the first Martian object—it was more a projectile than a rocket—lands in southern England about a hundred miles from London. Today, such an event would lead to an immediate news flash over TV and radio and the prompt arrival on the scene of the military, police, scientists, and the news media. It could not have been so in the England of 1897, and we therefore see the onset of the invasion as a much slower affair. At first only those in the surrounding villages are aware that some kind of strange celestial visitor has landed in the sand pits near them. It is only hours later that a few brief lines appear in some of the

London evening papers to the effect that a "meteorite" has plummeted to Earth in Surrey and dug itself a crater. No army units arrive, the police are not particularly interested, and only a few spectators eventually turn up to gape uncomprehendingly at "the thing from space." It is only many hours later when repulsive octopuslike creatures have emerged from the object and set up a devastating heat ray, which incinerates the Astronomer Royal, two other astronomers, and several foolhardy and overinquisitive spectators, that army units begin to arrive—engineer corps, some infantry, and several batteries of horse-drawn artillery. In the ensuing encounter with the Martians, most of the troops, the guns, and their crews are ruthlessly swept out of existence and the panic begins. But yet even now, because such communications as there are have been cut, the sprawling metropolis of London remains unaware of the terrible threat growing on its very doorstep. Wells tells of trains speeding by only a few miles from the stricken area. Slowly, we see the contagion spreading. Now there are rumors in London; railroads between the capital and the south coast either cease to operate or are seen to be carrying only immense siege guns and loads of heavily armed troops and marines. Rumor feeds on rumor. There are confused stories about "octopus men from Mars," which at first are treated mostly with scepticism and derision. But soon the reader senses the laughter growing more hollow. Something is undoubtedly amiss down there to the southwest. Tales begin to spread of refugees reaching the outer suburbs of the capital with strange and terrible tidings.

Slowly, almost imperceptibly, the dam begins to burst. Late on the Saturday evening of that fateful weekend the sound of faintly rumbling gunfire to the southwest is heard. The skyline is lit by flashes like summer lightning, and now strange patches of hitherto unknown red weed appear on the surface of the River Thames as it flows past London. Has this some connection with the mysterious happenings? And what about the occasional corpse? Yet in the very heart of the great metropolis life goes on much as usual. Theaters, music halls, and restaurants are packed; carriages and hansom cabs are doing a roaring trade; the barrel organs play as merrily as ever. In it all one senses a "night before Waterloo" atmosphere. So London plays; then goes to sleep. But in the early hours of Sunday morning all is dramatically changed. The battle is fought and irretrievably lost, the defending British armies destroyed. In that day and age there can be no vital news flash, no government announcement to reach the populace instantaneously. The news comes by "bush-telegraph," from street to street, from house to house. And though the threat facing the capital and its inhabitants is great by any standards, we see the terrible effect of rumor feeding and growing on little-known fact. Panic, tremendous and terrible, breaks out. Within a few hours the organized life of London and its environs is at an end. People have but one aim— to quit the metropolis—and, since the threat is apparently enveloping the city from south and west, escape is eastwards downriver to the sea and thence by steamer to the continent or north by railroad. But soon the railroads are unable to cope. Mob rule becomes the order of the day. We read of trains running one after the other with no attempt at timing or signaling, of coaches crammed full, with men, women, and children on the roofs and clinging to the sides. And so the Martians reach London and begin its systematic destruction with their terrible heat ray.

Those who reach the coast to the east find only brief respite. Wells speaks of vessels large and small standing out to sea loaded to the gunwales with terrified passengers who've paid exorbitant fares for the passage. And then, horror of horrors, the first of the great Martian tripod fighting machines is seen on the horizon. Engines throb; funnels belch smoke and sparks; sails billow as each craft puts on all possible speed. Now the battle squadrons of the British navy appear to protect the endangered shipping (after what has happened to the army one does not feel overoptimistic about their capacity to achieve this). The Martian fighting machines reach the coast and begin to wade out toward the fleeing ships. It is at this point in the story that the Martians suffer one of their few casualties. Suddenly, at prodigious speed, with foam curling and threshing from her bows and stern, appears the cruiser *Thunder Child* steering straight for the nearest Martian war machine. At almost point-blank range her great guns loose one tremendous salvo. Huge high-explosive shells burst on and around the Martian machine, which is rent asunder, collapsing into the water in a vast cloud of steam. For

Thunder Child the moment of triumph and glory is brief. Another Martian machine virtually sears her in two with its heat ray; her magazines explode in a mighty paroxysm and she is gone.

Here then is space war—or at least war brought on by an invasion from space—in all its naked horror. When the book was written, such an eventuality was seen as quite fantastic. Men had only just begun to fly—and most precariously at that. Aeronautics was the great dream of mankind, not astronautics. The book was fiction, and of the most fantastic kind. Certainly there were wars and rumors of wars—colonial skirmishes in Africa and Asia, small brushfire wars in the Balkans, and the already looming but still undiscernible holocaust of 1914. Aerial warfare was envisaged and its ultimate horrors already imagined. H. G. Wells was in on this, too. In his celebrated *War in the Air* we read of a great German air fleet attacking New York City with bombs in 1906. But space warfare was, like space travel, another dimension, a nightmare that could never come to pass.

Today, as the twentieth century begins to draw to its close, concepts are changing. Atomic bombs, dropped in anger, have already "taken out" two cities; aerial warfare during World War II completely devastated many cities in both Europe and Asia. Men have walked on the Moon and sent instrumented probes to Mars, Mercury, and Venus. The realization that the universe almost certainly contains an infinity of other worlds and probably other beings is securing an increasing acceptance among men of science. With that the link is forged. If representatives of our kind have trodden the surface of a nearby world, could not others, much more technologically advanced, reach out toward remoter worlds, such as our own? And since ours is a pleasant world with abundant water and a temperate climate, might it not be the envy of alien peoples whose own worlds are aging or dying? And so we come full circle to Wells and his *War of the Worlds*.

Now all this is not to suggest that mankind goes in daily fear of an onslaught from outer space. He is much too preoccupied with his own affairs and problems. Moreover, it is at best doubtful if more than a minute proportion of Earth's teeming millions even realize that other inhabited worlds are a possibility. How many appreciate that a star is merely another very remote sun—

or conversely that our all important sun is just another, very average star? And among those who are vaguely aware of such things, how many really care?

Nevertheless, there is an increasing awareness in scientific circles that Earth and our Solar System might not forever remain inviolate, that peculiar strangers in odd futuristic spacecraft might one day appear in our skies. No doubt our less inhibited "flying saucer" enthusiasts will claim they already have. In a very few instances the enthusiasts *might* just be right. If advanced aliens were to reach our world, their weapons would be equally well advanced; and so we have the prescription for an invasion from space.

An invasion of Earth by nonterrestrials is not, of course, space war in the absolute sense of the term. Space war proper is still to us the world of Flash Gordon and more recently the TV serial "Star Trek" and the film *Star Wars,* in which starships from opposing worlds do battle in deep space. Such events may already be occurring in deadly earnest in far-off corners of the galaxy. In the days of Flash Gordon and his friends, death rays and disintegrating beams belonged exclusively to them and their opponents, a fact for which most of us, deep down, were profoundly thankful. Though they still do not exist, the rapid onset and development of laser technology gives one the highly uncomfortable feeling that they are not perhaps all that far distant. Inevitably one wonders whether or not farseeing men, reading seventy years ago of the atomic bombs in Wells's *War in the Air,* pondered the advance of atomic knowledge and had similar thoughts. If any did and lived to see the year 1945, they had their answer. History, it is said, has a strange habit of repeating itself. We can only wait and see.

So far we have been thinking only in terms of space war between ourselves and aliens or between aliens and other aliens. But what of war fought in space between terrestrials and other terrestrials? Early conflicts on our world were fought only on land and on sea. Later ones, as we are all too well aware, were fought in the air, as well. Might not any future ones (perish the thought) be fought in space too—perhaps even exclusively so—as nations, superpowers, power blocs, and rival ideologies fight over mineral-rich or strategically placed planets, moons, and asteroids? Perhaps. But a mere half century ago or less the idea of walking upon the Moon or viewing the barren surface of Mars from over sixty million miles of empty space would have been seen as quite ridiculous, too. The fantasy of yesterday, be it good or bad, has a way of becoming the fact of tomorrow. The population of our already overcrowded planet is multiplying alarmingly, almost exponentially. Its resources are clearly and rapidly dwindling. We are beginning to need the resources of other planets. The moral, like the threat, becomes increasingly plain—stark might be the better term. Of course man should be able to arrange his affairs in such a way that he does not fight with his fellow man. Thus far he seems unable to do so. And therefore fleets may yet battle in space, every vessel of which was constructed by terrestrial hands and is manned by terrestrial crews. No doubt moralists will cry out that this must not be. If they were to say "should never be," it might be easier to agree. Knowing the frailties of mankind it would indeed be a sanguine person who would view the situation with confidence or optimism. Our world (and human nature) is as it is—not as we might like it to be. But this particular aspect really represents a digression, since our theme in these pages is not that of internecine war between the nations of this planet. It is, rather, the possibility and pattern of likely conflict between ourselves and alien races who at some future point in history might enter our solar space with hostile and aggressive intentions. "Science fiction," a sceptical reader might be tempted to murmur. In the interests of mankind we would sincerely hope him to be correct, but there can be no guarantee. The universe, just like our world, is also as it is—not as we would wish it to be. Very recently Sir Bernard Lovell, Director of the famous Jodrell Bank radio-astronomy observatory in England and one of the world's foremost radio-astronomers, gave this warning: "We must regard life in outer space as a real and potential danger. You have only to think about the problems of diminishing resources here on Earth to realize that alien civilizations may be combing the galaxy, looking for new resources or a new place to settle. They could want something we've got—and they could well have the ability to take it from us whether we liked it or not!"

2. AN EAR TO THE VOID

Whereabouts in our galaxy alien civilizations having highly advanced technoligical abilities might exist has been dealt with in a number of works, including one by the present writer. Over the past few years, however, research in this field has escalated so greatly that a measure of updating is definitely called for. It is by the location of such centers that we can pinpoint sources of possible danger to our world and its society. The distances involved, even if quite immense by stellar standards, should not necessarily be seen as some kind of broad, uncrossable moat behind which "Fortress Earth" can lie forever inviolate. Civilizations several millennia in advance of our own could, by virtue of undreamed of technology or swift passage through unsuspected dimensions (unsuspected, that is, by us), have rendered that selfsame moat little more than an easily traversable ditch.

The pioneer listening program for extraterrestrial and extrasolar radio signals was, of course, the now almost legendary "Project Ozma," conducted by Dr. Frank Drake and his colleagues at the National Radio Astronomy Observatory, Green Bank, West Virginia, during the summer of 1960. A 85-foot radio-telescope was employed, and for a total of 150 hours two "close" solar-type stars, Tau Ceti and Epsilon Indi, were monitored for possible radio transmissions of artificial origin. Though the search proved fruitless, little actual disappointment was felt since it was realized from the outset that a mere two stars hardly consititued a very comprehensive program. Moreover, it was appreciated that the radio-telescope used was not really adequate for such an undertaking. "Project Ozma" represented merely the exciting genesis in the new field of exobiology and its first practical expression.

Today, man's quest to intercept extraterrestrial signals advances on a much broader front, using wavelengths covering a wider portion of the electromagnetic spectrum (radio to ultraviolet) with targets ranging from "nearby" stars to other galaxies. The fact that such interest and escalation in activity is taking place reflects the growing realization that intelligent life in the galaxy—and indeed in the universe as a whole—is probably very widespread. Study groups of scientists such as those attending the Soviet-American Congress on Extra-Terrestrial Life (CETI) at Byurakan, Armenia, in 1971, have suggested that there could be as many as *one million* highly advanced civilizations in our galaxy at the present time. The subject has therefore been divorced to a large extent from its earlier and almost exclusive science fiction connotations. That in itself is no bad thing.

Estimates made take into account the probable frequency of planetary systems, and the likelihood that life can and will arise on appropriate planets. And though a civilization such as our own, or one in advance of it, may develop around only one star in a hundred thousand, our galaxy, nevertheless, has more than sufficient stars (around one hundred thousand million) to render a protracted search well worthwhile.

So rapid has been the pace of technological advance in this field recently that five minutes of star-monitoring with the present Green Bank 140-foot radio-telescope is now reckoned equivalent fo *four days* using the original "Ozma" 85-foot instrument and associated circuits.

In 1971, eleven years after "Project Ozma," Tau Ceti was examined again at Green Bank, but on this occasion the giant new 300-foot "dish" was used. Also included in this program were the stars Epsilon Eridani and 61 Cygni. The following year the 140-foot instrument was brought into use during a 21-cm wavelength search for emission from ten nearby stars (in stellar astronomy "nearby" is very much a relative term) including once again Tau Ceti, Epsilon Eridani and also on this occasion, Barnard's Star, around which a planet, or planets, was first indicated by Peter Van de Kamp in 1963. Barnard's Star, incidentally, is also the second closest star to our own sun. The official verdict at the close of this search ran, "No signals of the order of 1 megawatt of power were then being transmitted from a 300-foot diameter radio-telescope orbiting any of the ten stars examined." It is important to note that this brief and rather laconic report does not imply that no signals were possible—only that none had been detected with the equipment used. Sensitive and ultrasophisticated alien radio-telescopes directed at our own Solar System might produce similar apparently negative results. This would not imply an absence of advanced life in the Solar System. It would merely indicate that the alien equipment had been unable to detect our indigenous radio and radar signals, though a different result might have been forthcoming had we deliberately been beaming a powerful signal toward their planet.

Precisely which frequencies alien radio astronomers might elect to announce their existence to the universe or to communicate with other galactic communities is a question to which our own radio astronomers have, over the past two decades, devoted considerable thought. The wavelength on which the great clouds of hydrogen emit, 21 cms, has for long seemed the most appropriate because most radio-astronomy receivers in the galaxy will at one time or another be tuned to it. The objection to this is, of course, the rather obvious one: "Hydrogen noise" may simply mask intelligent radio signals and for that very reason 21 cms may be the wavelength that should *not* be used! For some time now this has seemed to the present writer distinctly and increasingly plausible.

Recently, the discovery of the molecules of certain elements and compounds in the vast interstellar spaces has opened up the possibility of utilizing the frequencies on which these emit. One such molecule is, oddly enough, that of water, which emits radiation on a wavelength of 1.35 cms. This has been suggested, perhaps a little facetiously, as a highly appropriate wavelength for water-based beings. No one in the past ever seemed to harbor analogous ideas regarding hydrogen-breathers on the 21-cm channel.

At present the first-ever search of the "water wavelength" (1.35 cms) is being carried out in Canada using the Algonquin radio-telescope. This search was instituted in 1974 with a program embracing thirteen stars. It is hoped eventually to extend the search to include 500 "nearby" stars similar in most repsects to our own sun.

Probably the most comprehensive search of nearby stars in the galaxy was just recently (1977) completed at the National Radio Astronomy Observatory, Green Bank, by Benjamin Zuckerman and Patrick Palmer, using the 300-foot partially steerable antenna and the 140-foot equatorially mounted, fully steerable antenna. The search, given the title "Ozma 2," began in 1972. Initially, the program was scheduled to last for two years but, in fact, it ran considerably longer. Instead of terminating in 1974, observations continued thereafter on a sporadic basis until December 1976. The National Radio Observatory authorities, reflecting fully the rapidly changing outlook in such matters, remained receptive to requests by Zuckerman and Palmer for additional observing time. This valuable concession enabled them to include more stars in their search. In all, approximately 700 stars were examined during the run of "Ozma 2" out to a distance of 65 light-years. Target stars were selected from the nearby stellar population on the basis of luminosity and spectral class. Prime targets were main-sequence stars in classes F5 to K5—the astronomical shorthand to denote stars akin to, or rather similar to, the Sun. Observations were again carried out on the 21-cm "hydrogen" wavelength; each of 384 separate radio receivers was tuned to a slightly different wavelength, but centered on 21 cms.

Zuckerman and Palmer estimated that they could take data *10 million times* faster than in the original "Ozma" project of 1960, and hoped to detect a civilization equivalent to our own (i.e., a 40-megawatt transmitter beaming through a 100-meter telescope—the contemporary technological criterion for terrestrial society) around one or more of the "target" stars.

The additional observing time allocated to Zuckerman and

Palmer during 1975 was used to survey 130 stars more using the 140-foot radio-telescope. They also used this instrument to carry out a more intensive examination of ten of the stars previously scrutinized. These stars, according to Zuckerman, showed "glitches," i.e., suspicious emissions. Some of these glitches were eventually attributed to aircraft. The others, however, have not so far been explained and remain a mystery. Is it possible, just remotely possible, that here we have the first tangible indications that we are not alone in the universe?

"Ozma 2" has already spawned other research in this field. Tom Clark of NASA's Goddard Space Flight Center and Jill Tartar of NASA's Ames Research Center have since undertaken a small-scale search project at Green Bank employing the 300-foot antenna. They are also using a receiver with extremely narrow bandwidth (only a few cycles per second compared to the 4,000 cycles per second bandwidth of the receivers used in "Ozma 2"). Though this narrowing of bandwidth increases the amount of time required to search a given segment of the radio spectrum, it markedly enhances the sensitivity to any very faint signal within the portion searched. In fact, this search by Jill Tartar and Tom Clark is reckoned to be *a thousand times more sensitive* than "Ozma 2."

On completion of "Ozma 2" Dr. Palmer said that, in his view, a successful search might easily take decades, if not centuries. "We have no right to expect success within our lifetime," he stated. This, of course, is simply a reiteration of the obvious. The galaxy probably teems with life, much of it highly advanced, but the fact remains that immense distance is a difficult parameter to overcome. Indigenous alien radio emissions are almost certainly beyond the ability of our relatively crude equipment to detect, while true interstellar signals may come our way only briefly and rarely—as yet. If we are not in the line of the beam we receive nothing, even though the star concerned is relatively close to us. Conversely, a beam reaching us from an immense distance could have become so attenuated that we are unable to detect its presence. Only if a "close" star beams directly at us are we likely to be aware of it. Clearly, the odds are against us. In the context of this book, then, it must be admitted that we could in the future be invaded by beings from a planet or a star from which we had heard precisely nothing. In other words, pinpointing stars with advanced technological societies cannot be reckoned as an infallible first line of defense. On the other hand, to know positively that, for example, only six light-years distant there exists a civilization of great power and potential might, at least in time, persuade the world's rulers that the impossible is just possible, that the unthinkable must become thinkable.

Dr. Palmer feels that the now traditional technique of selecting "close" stars and monitoring them at a wavelength of 21 cms has been exhaustively applied and that the time has come for the adoption of more sophisticated techniques. The efficiency of any search, he suggests, could be greatly enhanced by the development of radio receivers having more sensitivity over a wider bandwidth than those we now have. He adds, "If in the next decade or so sufficient progress were to be made in this area the time required to find our needle in the cosmic haystack could be dramatically reduced." It would truly be ironic if, during that time, the needle was found and grievously injured us.

Dr. Robert S. Dixon of Ohio State University has, since December 1973, been conducting continuous sweeps of the sky in search of possible 21-cm emissions from advanced galactic communities. The technique he uses is to tune to the precise frequency that would be received from a 21-cm transmitter located at the *center* of the galaxy since he believes that other cosmic civilizations would modify their signals so as to eliminate Doppler effects caused by the motions of stars around the galaxy. He reports, "So far no objects have been discovered which we believe are emitting intelligent signals, *within the constraints of our observations.*"

Another improvement scheduled to be made soon at Ohio State is an enhancement to the computer's capability to automate the search. Originally the graph produced by the receiver had to be analyzed manually. The analysis was both tedious and time-consuming and also, of course, open to human error. The computer has now been programmed to analyze the output of each receiver, rejecting everything except recognizable signals.

The Ohio State program has now been in full-time, 24-hour operation for over three years. In that time no extraterrestrial signals of intelligent origin have been detected, though one or two false alarms did occur. The report does add, however, that "many objects worthy of reexamination have been detected, and this reexamination process is under way at the present time." Inevitably, the question arises, "Where is everybody?" It is widely accepted among radio-astronomers engaged in these searches that the lack of tangible results so far should cause absolutely no surprise. We may require, they claim, to search out as far as 1,000 light-years or so before intercepting the first extraterrestrial intelligence. This involves the examination of perhaps 100,000 stars. So far in all these searches about 1,000 have been covered—and that perhaps only with inadequate equipment. The report from the Ohio State team does contain, however, an item that is both timely and intriguing. "Out to a distance of 1,000 light-years," it reads, "there is an average of three F, G, and K type stars in the beam of the radio-telescope at all times." It is such stars that are considered most likely to have spawned life-bearing planets.

The threat to us of an invasion by alien beings from the worlds of other suns or of our becoming involved in a star war may seem very remote indeed. If we consider that the very nearest star to us, Alpha Centauri, is 4.3 light-years away, the distance seems for all time unbridgeable. Alpha Centauri is very roughly 25 million, million miles away, and a space vehicle traveling at the incredible speed of one million miles per hour would take around *2,700 years* to complete the journey. The moat we spoke of earlier seems more like a limitless ocean in these circumstances. This is perhaps the feeling we should most guard against. To aliens who might, for all we know, have developed the technique of rupturing the "skin" of the space-time continuum, thereby to traverse a time-shrinking extra dimension, the Sun and its family of planets could lie relatively near. Beware the unsuspected, unseen tunnel in space.

The serious search for extraterrestrial civilizations among the stars has been steadily gathering momentum since 1960. More-over, it is one that appears to be transcending the ideological schism that divides our world at the present time. We find that the Soviet Academy of Sciences in March 1974, approved a ten- to fifteen-year search program split into two parts—CETI 1 from 1975 to 1985 and an overlapping CETI 2 from 1980 to 1990. The former includes all sky monitoring from eight ground stations and two satellites (as well as studies of "nearby" galaxies). CETI 2 involves the use of more sophisticated equipment for the examination of selected stars. Ultimate CETI plans by the Soviets envisage the monitoring of all appropriate stars up to 100 light-years distant and eventually to 1,000 light-years.

It has frequently been pointed out that if all galactic civilizations, including our own, were merely to adopt a passive role, i.e., listening but not transmitting, then listening would simply become a pointless exercise, since plainly there would be nothing to listen to. It seems most unlikely, however, that all the many galactic civilizations we believe may exist have been struck electronically dumb. Admittedly, listening is a prospect that, in the first instance, is the more fascinating, since meaningful results could, with luck, be achieved anytime or at least within a measurable future. Signals transmitted by us will, on the other hand, not be received by aliens for several years, perhaps not even within our lifetime.

We should not, of course, be unmindful of the security aspect. The theme of this book is, after all, that of possible invasions of our world and Solar System by intelligent beings from worlds around other stars. Why, it might well be asked, should we increase the danger by drawing the attention of these aliens to the existence of our pleasant world and its civilization? Unfortunately, it is probably a fairly safe premise that aliens with the technical prowess to reach us will long have had the ability to detect our indigenous radio signals. During the last two decades a very extensive radar surveillance has been in existence over our planet in the interests of national security. We have also sent ultra-high frequency pulsed radar signals to the Moon and radio-controlled probes to Mars and beyond. Might not these have been detected by intelligent beings on worlds around some of the nearest stars? It would be a sanguine person who was willing to give a categorical negative to that question!

A number of civilizations within our galaxy could, at this very moment, be using a highly efficient "lighthouse beam" system either in the interests of galactic research—or as a ruse to tempt a hitherto silent society into revealing its presence. They would be sweeping the beam along the *plane* of the galaxy, where most stars lie. It is reckoned that a technological ability not greatly in excess of our own at the present time could achieve this. (Fig. 1). "Dishes" similar in size to that of the Jodrell Bank radio-telescope in England would probably suffice and these would not even have to be fully steerable. Using 4 megawatts of power and a bandwidth of one Herz (one cycle), such a beam could briefly play upon us from distances of several hundred light-years. A few brief bleeps from us in return and "they" know we are there—and also perhaps *where*. To a highly advanced society would even these return bleeps be necessary? Might not their beam be capable of remote sensing in a way somewhat analogous to techniques that we have already developed with respect to the Moon and Mars? These possibilities are dealt with in the following chapter.

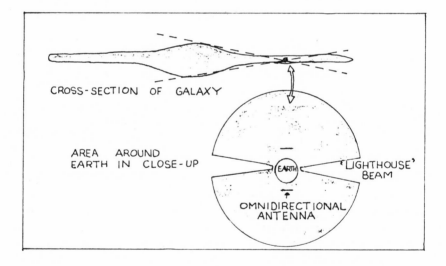

Figure 1

Here then in our present context is what, despite its obvious shortcomings, must be reckoned our first line of defense—listening for possible intelligent interstellar signals that would indicate that at least *one* other advanced civilization and technology exists within the galaxy. By inference there would almost certainly be others—perhaps an almost unlimited number. A radio beam reaching our Solar System from the depths of interstellar space might have purposes other than just mere communication. It could have sensing potential enabling its originators to learn intimate details of distant worlds, their terrain, atmosphere, gravity, temperatures, and if inhabited, details concerning their civilization and defenses.

In looking for evidence of such beams we need not necessarily think only in terms of their originating from the star worlds themselves. They might very well originate from points in space much closer to us—points occupied by roving alien cruisers selecting possible planets for the next wave of alien colonization.

In all of this we must, however, always remember that radio waves, though they move with the incredible velocity of light itself (186,000 miles per second), still take years to traverse the interstellar gulfs. If next week we were to discover that a highly advanced alien society existed on a world of the star Tau Ceti and that its members were bent on an invasion of the third planet of the Solar System (ours), the information might not be of very much military value to us. The message we intercepted would have been sent out eleven years ago (since Tau Ceti is eleven light-years distant), and therefore the invasion force would have to be regarded as well on its way. And if by any chance it employed some relativistic transit technique or travel through another dimension, then it has probably already arrived!

Nevertheless, in the theme of "Space Weapons/Space War," we may have to see the radio-telescope as something of a weapon, a device that could give us the first hint of a potential or impending space war.

Sooner or later too, we ourselves will begin to develop an interstellar travel potential. If by then Earth and the Solar System still remain inviolate, it would be as well to know where other advanced galactic communities exist so that we can go in the opposite direction. There seems little future in blundering into someone else's star war.

3. THE UNSEEN EYE

In the preceding chapter we had a brief look at the possibilities of locating advanced galactic communities by electronic means, i.e., by detection of any radio beams such communities might be transmitting in our direction. We mentioned that such beams might conceivably have a purpose other than to attract our attention. Let us now explore this aspect in a little more detail.

In any military operation, especially an invasion, it is not a bad idea to have some notion of the strength of the defending forces, their deployment, and the nature of their fortifications and weapons. An interplanetary war would be unlikely to alter materially this state of affairs, though intelligent alien beings capable of rapid transit through interstellar space would, we must imagine, be in a position to squash quickly and most effectively any opposition offered to them by the peoples of our planet. Nevertheless, despite an enormous techno-military advantage, preliminary reconnaissance by the aliens would probably still be regarded as highly desirable. Just how could this be achieved? One possibility is that of remote electronic sensing instituted either from the home planet(s) of the potential invaders or carried out by reconnaissance spacecraft lurking within the outer perimeter of the Solar System. The latter seems the more likely.

Already we on Earth have achieved a considerable measure of breakthrough in this respect. What we have achieved, it might justifiably be argued, bears no relation whatsoever to similar sensing over vast interstellar distances. Undoubtedly true—but what might be our own powers in this field, say a thousand years from now? Prior to 1946 we had no such abilities whatever. Today we have been able to secure a considerable amount of valuable data concerning the surface of the Moon. Such data, it must be pointed out, was gained from Earth and had nothing to do with subsequent lunar landings. What then might we justifiably expect from vastly superior intelligent beings in a nearby star system or aboard a highly sophisticated space vessel lurking on the outer fringes of our Sun's family of worlds?

First of all, let us examine the development of this technique as it applies to ourselves and then try to consider what extrapolations might be expected from another galactic community a few light-years from us and several millennia ahead.

Following the close of World War II in 1945, workers in Hungary, and independently at the U.S. Army Signal Corps Laboratory, succeeded in detecting radar echoes from the Moon. This represented a very considerable breakthrough in astronomical science for two specific reasons:

1. For the very first time ever, research could be carried out on an *active* instead of on a purely *passive* basis. Thanks to radar the Moon could now be "illuminated" by specific forms of electromagnetic energy and the radiation scattered back by the lunar surface measured and interpreted. All previous work in this field had involved analysis of whatever *natural* radiation was available from the Moon.

2. *Optical* methods of examination previously employed had provided data only about the extreme outer surface of the Moon. Now, by virtue of radar, it was possible to determine something of the properties of material lying a short distance *beneath* the lunar surface.

Two kinds of radar technique are currently in use. They are known as short pulse and delay Doppler, respectively. It must be pointed out that by now they have been employed in the mapping of Mercury, Venus, and Mars, as well as in investigations of the lunar terrain. Let us have a brief look at each.

Short Pulse Measurements

Suppose the Moon is to be "illuminated" by a radar beam antenna that transmits a burst of energy having very short duration (Fig. 2). After a transit time of approximately 2½ seconds (since the Moon lies 1¼ light seconds from Earth) a radar echo is received. It should be noted that because the Moon is nearly spherical the echo takes slightly longer to return from the most distant points on the lunar limb than from the point on its surface nearest to the transmitter. (The latter is known as the subradar point.) The difference in time between an echo from a limb point and that from the subradar point is 11.6 microseconds. The echo from the rim, as reference to the diagram will show, represents the integrated (i.e., total) power from an annulus, or ring, on the Moon's surface (Fig. 2).

Because the pulse of energy is so brief, this annulus is narrow in width. Nevertheless, even with a pulse of only a few microseconds' duration the area of the annulus is of the order of a few thousand square miles. In other words, the technique can only provide information averaged over a fairly considerable area.

One of the great advantages of radar is that it enables astronomers to transmit pulses of radiation having intrinsically different properties and to examine the different effects produced. It has been found that the use of *polarized* radar signals represents the most profitable form of research. Radar signals are a form of electromagnetic radiation just as light is. In this case, however, the frequency of the radiation is of a much lower order (i.e., the wavelength is considerably longer). Such radiation can therefore be polarized in an analogous way.

A polarized radar signal transmitted to the Moon is generally scattered back in the form of two precise components, one being polarized and the other depolarized. It is not really necessary that we go into the physical reasons for this here. What *is* important is the result and its implications. (The ratio between the polarized and unpolarized components varies from about 1,000 to 1 at the subradar point to around 2 to 1 at the limbs). Both, however, decrease in strength as the angle of incidence decreases (Fig. 2).

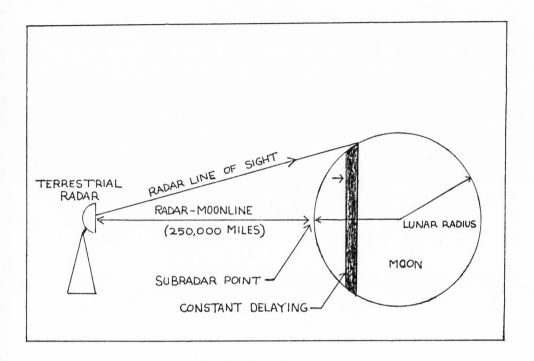

Figure 2

An echo of this sort is termed specular, that is, it is mirrorlike in nature and is due—and this is the really important point—to *scattering by a large number of smooth facets many wavelengths in size.* The great significance of this lies in the fact that a relationship exists between the wavelength used and the "facets" doing the scattering. In this instance the wavelength employed was 23 centimeters, which means that (for physical reasons which we will not go into here) the "facets" must be one or two meters in size. Another fact that can be gleaned is that the *average slopes* of the lunar surface are of the order of 5 to 10 degrees. What then are these mysterious little objects which, for the time being, we have termed "facets"? The size and angle of the slopes suggests that they are, in fact, *small craterlets.* Craters having a diameter of 1 to 2 meters are, of course, quite invisible in even the largest of Earth-based telescopes, which, so far as craters are concerned, are incapable of resolving anything less than 500 meters in diameter.

Nor is this all the hitherto unknown data which radar technique makes available. The *total amount of backscattered power* also provides information about the *electrical properties* of lunar surface rocks, which are found to be consistent with those of terrestrial rock dusts.

It has also been found that longer wavelength radar pulses will penetrate *beneath* the lunar surface, though not to any substantial depth. Their use has revealed the fact that just beneath the lunar surface are a large number of buried rocks.

Thus, without stirring from the surface of our planet, we were able to gain a little insight into some of the more intimate geological details of the topography and immediate subsurface of another world a quarter of a million miles distant.

Delay Doppler Technique

One disadvantage of the short pulse technique just described is that it provides integrated data over the *entire area* of an annulus on the Moon's surface. It is unable to provide data over any specific small area. The delay Doppler technique, on the other hand, permits quite small parts of the lunar surface to be examined. In pre-Apollo days this was a most decided asset.

The technicalities are somewhat involved, but in essence what happens is that an echo annulus at a constant range from Earth is formed on the Moon's surface as before. This is intersected at two points (Fig. 3) by another, due in this instance to Doppler shift. The *small areas where the two intersect* can then be studied fairly objectively. The returned radar echoes from these small areas are analyzed in the same way as short pulse echoes. Considerable information can be gained as a result of this technique that confirms the loose, rubbly nature of the lunar surface and shows that it has radar properties similar to a dry, dusty desert having particles with an average grain size of around one hundredth of a centimeter. The technique also shows that the surface material has electrical properties akin to those of terrestrial volcanic rocks and that the lunar surface is probably layered in places.

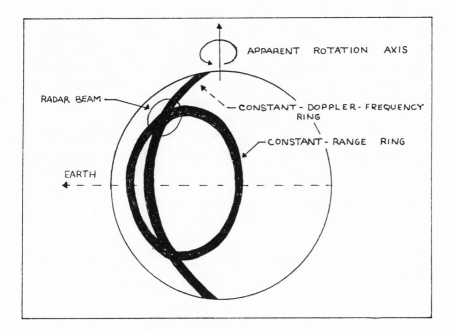

Figure 3

At this point it might be a good idea to see what relevance all this has to our present theme. We have seen how a certain amount of intimate detail pertaining to the Moon's surface and immediate subsurface has been obtained using radio equipment and nothing more, from the Earth, a quarter of a million miles distant. Most people would agree that this represents a quite remarkable feat, though it is one that has been much overshadowed by the unmanned and later the manned lunar landings—and, of course, also by the recent and very highly successful "Viking" Mars landers. Nevertheless, its true import should not be overlooked. Space travel is neither cheap nor easy, yet here is a method that can provide a certain amount of fairly intimate detail about other worlds without the necessity of having to stir a single foot from the surface of our own.

What relevance, though, the reader may still ask, can this have in relation to interstellar distance? We may indeed be able to glean a modicum of knowledge concerning the surface of relatively close sister planets, but any such technique would prove totally unavailing in the case of planets orbiting other stars, planets that we cannot even see. This is undoubtedly true. Once again, however, we must bear in mind that all important parameter: Though some civilizations in the galaxy will be behind us technologically and others our equal, many will be far ahead, some perhaps to the extent of several millennia and more. At present we are many millennia ahead of Stone Age man who was, we can reasonably assume, unable to probe the Moon's surface by means of radar. The intervening millennia did much to develop man's intellectual powers. The parallel is clear. The terrestrial civilization millennia from now (assuming we have not destroyed ourselves in the interval by some crass piece of nuclear folly) will doubtless look back on our original radar exploration of nearby worlds as something very primitive indeed. In the same way, advances made in this field by contemporaneous but highly technological cosmic societies will have rendered their techniques equally far ahead—perhaps even to the extent of being able to study our world by some form of remote and sophisticated electronic sensing. Dr. Carl Sagan puts the case very well when he says: "Civilizations hundreds or thousands or millions of years beyond us should have sciences and technologies so far beyond our present capabilities as to be indistinguishable from magic. It is not that what they can do violates the laws of physics; it is that we will not understand how they are able to use the laws of physics to do what they can do."

Sophisticated electronic sensing by alien beings represents a monstrous extrapolation of our own abilities in this field. By such means some might easily have established the fact that the star we call the Sun has a retinue of nine planets; they might even be aware of the existence of one or two beyond Pluto still unknown to us. More precise beams might then start a cosmo-clinical examination of each. Eventually it would be the turn of Earth. Here is a planet neither too near nor too far from its central sun. "This is good," they say, realizing that their own world is dying and aware of the urgent need to find an alternative abode. They sense great oceans and masses of cloud. "Excellent," they exclaim, thinking how dry and parched their own world is becoming. And then perhaps, just as we sensed small craterlets on the Moon and boulders on and beneath its surface, they detect not only continents, mountains, lakes, and rivers, but also the artifacts of man. Large objects are apparently traversing these oceans; great metal birds pass over both them and the continents. "Not quite so good," they say, "a primitive technological civilization exists here. Still, no matter, we have the means of easily overcoming or removing that!"

It could be asked that, primitive though we may be, surely we in turn could detect these whispering, probing, feeling rays focused on us by races several light-years distant. It is a good question, a moot point, a seemingly very valid inquiry. We should, however, recall the words of Carl Sagan: "What is apparently near *magic* to us is orthodox *science* to 'them.'" The electronic beams might simply be too weak for detection by our relatively crude equipment. That is just one possibility. Alternatively, they could be using a part of the electromagnetic spectrum unavailable to our contemporary technology. We can be reasonably certain that an advanced cosmic civilization with planetary aggression or star war in mind will seek to ensure that the worlds over which they are maintaining surveillance will not easily be rendered aware of the fact. In any world or in any galaxy that would hardly be good military practice.

What was done with respect to the lunar surface from Earth itself was later amplified by radar examination of the Moon from orbiting spacecraft. Profiles of the Moon's topography could then be recorded in a graphical form (Fig. 4). The low-lying "maria" can be clearly identified, as can the so-called lunar highlands. Note, for example, the clear dip in the profile due to an as yet unnamed crater.

In our present context there is something of a parallel to this even if it is a greatly extrapolated one. Should distance of interstellar dimensions prove something of a hindrance to advanced cosmic civilizations, consider the possibility of what could be done by their starships on reconnaissance missions. We assume that such vessels would not come into Earth orbit to carry out such missions, though some of our less inhibited "flying saucer"

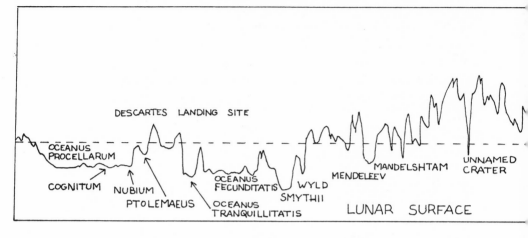

Figure 4

friends will no doubt disagree. Peoples with the technical prowess and expertise to reach the Solar System from the stars might well be able to achieve all that is necessary by carrying out their surveillance from somewhere, say, between the orbits of Saturn and Uranus. (Oddly enough, just a week or so after these words were written, a tiny object was discovered by Charles Kowal of the Mount Palomar Observatory in California. The object does indeed lie between the orbits of Saturn and Uranus, but its precise status is still something of a mystery. It is certainly not a comet and its dimensions are rather on the small side for it to be a planet. With an estimated diameter of between 100 and 400 miles, however, it seems unlikely to be an alien spy ship. At least with such dimensions we must sincerely hope not! It is in all probability a wayward moon of Saturn or Uranus that has decided to go its own cosmic way or the first to be discovered of a hitherto unsuspected outer asteroid belt. Its appearance at this time, so far as these pages are concerned, is merely a coincidence—a fortuitous one perhaps from the author's point of view, much less for humanity's if it should turn out to be anything less than a natural object.)

By merely orbiting the Moon we were able also to determine the nature of the Moon's magnetic field and, by use of an orbiting gamma-ray spectrometer, to determine the distribution of natural lunar radioactivity. X-ray flux from the lunar surface was also monitored from lunar orbit, thereby making it possible to determine the aluminium/silica and magnesium/silica ratios of the rocks, information of considerable geological and geo-historical value. Similar sensing techniques are already being carried out by Mars orbiters, enabling scientists to build a much clearer picture of conditions on the legendary red planet. It seems hardly unreasonable to assume that if we on Earth have been able to achieve all this a mere two decades or so after the advent of the world's first artificial satellite, galactic communities thousands of years older and more technologically adept than we would be able to achieve results of an infinitely more advanced and sophisticated nature. Has an alien probe somewhere out there in the frigid vastness beyond Mars already sensed the infrared radiation emanating from terrestrial cities and power stations and gone on to pinpoint their positions on a master plan of our world? Have they already detected the radio-activity resulting from nuclear tests and reached the conclusion that the society facing them is a little less primitive—and liable to be a little more troublesome—than they first imagined? Have they noted the early spaceships of Earth blazing their trails to the Moon, Mars, Venus, and beyond, and in so doing recalled their own long past cosmic genesis?

The night has eyes. This old saying may be a lot truer than we imagine. And the eyes may be very strange eyes—and a lot more discerning than we care to believe.

4. HOSTILES APPROACHING

In the realm of science fiction the detection of a hostile or potentially hostile space fleet, even a single starship, is generally seen as a fairly straightforward operation. And indeed, living as we do in a world of apparently highly sophisticated radar, we might seem to have no reason whatsoever to believe otherwise. There are, however, two pertinent facts that should be placed before the reader: Space is infinitely vast (in the truest sense) and in comparison a starship, mighty and imposing though it may be, is infinitely minute (literally). Such facts may seem obvious enough but frequently it is the obvious that is over-looked or disregarded.

In sci-fi novels and motion pictures the hostiles rarely, if ever, seem to escape detection. When finally they sweep in toward Earth, almost invariably the defending terrestrial spaceships are ready and waiting. Since good is generally expected to prevail over evil in the normal space opera, it might be rather awkward for author/script-writer/producer if the hostiles were capable of making an undetected approach, which enables them to surprise and defeat the powers of good.

In this chapter our aim will be to make a much more realistic appraisal of the situation. Should a star war ever occur in which our own familiar Solar System became involved, we would be confronted by a highly dangerous and potentially lethal state of affairs. It would, without a shred of doubt, represent the greatest menace our world and its peoples have ever been called upon to face. In a sense, the menace begins to take positive shape on the day our radio—astronomers finally confirm what so many of us already strongly suspect—that other intelligent beings exist in our galaxy. It would become more pressing were we to discover that such galactic communities inhabited planets orbiting stars only a few light-years distant from us—and even more so should it become apparent that such beings had already developed an interstellar travel capability. The menace would become potentially acute if it turned out their worlds were dying, growing short of essential resources, or alternatively that the inhabitants were by nature aggressive and ambitious. Earth had its Genghis Khans, its Attilas, its Hitlers in the prespace age. On certain other worlds we could never be sure that the position might not be reversed.

At the present time this may all seem highly (and safely) sci-fi—but then, a mere quarter of a century ago, so did a manned descent upon the Moon or the use of remotely controlled probes to show us the surface of Mars. The facts that exobiology is now being regarded as a serious scientific discipline and that electronic searches are already under way to detect other galactic communities show rather clearly that the concept of extraterrestrial intelligent life is being taken with increasing seriousness. And, after all, why not? The peculiar notion that throughout all the immensity of space and time we, and we alone, the rather pitiful creatures of a very small planet of a small average star in a quite average and undistinguished galaxy, are unique, is not only ridiculous; it carries faint undertones of megalomania.

So just as today the free nations here on Earth must zealously protect themselves from predatory totalitarian powers and ideologies, it may at some time in the future become vitally important, perhaps mandatory, to seek means of defense against roving galactic predators. In a populated galaxy there must exist the possibility of creeping, gradually encroaching colonization of suitable planetary systems by highly advanced races from others—a menacing wave, which, for all we know, may even now be moving remorselessly and silently toward our own Solar System. We could so easily remain in total ignorance of the danger until it was too late. The basic reason for creeping colonization, the surreptitious extension of alien dominion, could be a very simple one—the fact of population increase. Ultimately that increase becomes exponential. As it does, resources become increasingly strained, pollution steadily greater. Concurrently technology is advancing and in time a space potential develops. The reader need look no further than our own planet for proof of all this. Why then should things be different elsewhere? With the ultimate advent of *interstellar* capability a totally new and vital parameter is introduced. The bonds, the shackles, are broken and the teeming surplus, the refuse of the shores of other planets, can reach out and grasp new worlds. From dying planets, that grasp could be a strong and very desperate one. In time history repeats itself and yet more worlds become necessary. Should these be a bit more remote, have no doubt that advancing technology will provide the answers. The threat then, so far as we are concerned, could be very real indeed—and one that need not lie centuries or millennia hence, either.

For an effective defense it is necessary to have the right weapons, the right techniques, and the right strategies, but even these are of very doubtful value if the enemy possesses the ability to arrive and strike unexpectedly and unannounced. There are several instances of just this very thing in our own experience here on Earth, perhaps the most recent and blatant being the treacherous attack on Pearl Harbor by the Japanese on December 7, 1941. The equipment and personnel of the attacking force were not superior to those of their American counterparts. Indeed, the very opposite may well have been true (as events were soon to show, this was almost certainly the case). The overwhelming advantage lay with the Japanese, not because they were better, but simply because they were able to approach the Hawaiian island of Oahu that fateful Sunday morning undetected and from a totally unexpected direction. The rest is history. But had the military authorities received the necessary warning, it is a safe bet that the U.S. Navy, Army, and Air Corps would have given the marauders a highly unpleasant and very hot reception.

Now, whether in a star war the weapons and techniques of

Earth would prove adequate against the sophisticated and highly advanced technology of the invaders is, at best, a moot point; but if they were to have any chance at all, detection of the attacking forces at an early stage would be vital. Indeed, detection of the first signs of a hostile buildup deep in interstellar space would be highly desirable, though probably impossible to attain.

So, then, to the problems inherent in the detection of alien spacecraft approaching our world from beyond the Solar System. As a preliminary it seems safe to state that if interstellar travel is possible, we will most assuredly not be the first race within the galaxy to achieve the distinction. When we recall the ages of many appropriate stars in relation to that of the Sun, we realize that many alien peoples will inevitably have traveled farther down the highways of time and history. In the context of our theme then, we must see ourselves not as the potential invaders but as the potentially invaded. If our own world and Solar System do manage to remain inviolate throughout the centuries ahead, and we do not remove ourselves from the scene by some act of unmitigated folly (of which at the present time we are well capable) then supposedly in time the role and mantle of potential invader could descend upon us. This raises many difficult and highly ethical questions. If by then our planet had become so overcrowded, polluted, denuded, and generally unlivable, it is difficult to imagine our stressing ethics too much on discovering that a new, fresh, and unspoiled world lay within our grasp. Bad luck if that world contained an indigenous, more primitive people—*their* bad luck, not *ours!*

The prevailing assumption at present is that high energy spacecraft (i.e., starships) in our corner of the galaxy could be easily detected. It has, in fact, been estimated that such a spacecraft, traveling at around 13 percent of the speed of light, i.e., over 24,000 miles per *second* would, were it careless or unfortunate enough to collide with a small asteroid, release energy amounting to seven times that released when the island volcano of Krakatoa blew itself into a volcanic Valhalla in 1883. (For the record, such an explosion amounts to an energy release of 7×10^{19} joules.) The effects of a collision in space on this scale would hardly go undetected. Unfortunately, we cannot assume that all starships approaching our world with hostile intent are going to collide catastrophically with conveniently placed asteroids.

There are three basic techniques whereby, it is believed, a starship could be detected:

1. By virtue of energy emitted, reflected, or obscured by it.
2. By virtue of energy emitted from the engine or exhaust when an active propulsion unit is being employed (i.e., during periods when the power units are actually propelling).
3. By reason of volatiles, fragments, or artifacts released during flight.

One technique not included is detection of radio signals emitted by the spacecraft. We assume that like naval ships during World War II, the ships of spacefleets in a star war would be blacked out and keep strict radio silence on approaching an enemy.

As a preliminary, it is desirable that the term *starship* be examined. In broad terms it might be said that whereas all starships must by definition be spaceships, the converse does not necessarily hold. Spacecraft such as we employ today and even those which in the foreseeable future will ply between our planet and environmentalized colonies on the Moon and Mars, cannot be, or ever become, starships. (In a strictly technical sense they might, assuming their respective crews could arrange to increase their life spans by many thousands of years, at the same time dispensing with such frivolities as breathing, eating, and drinking.) The starship proper is a super spaceship capable of sustained flight at relativistic velocities because of the vast interstellar distances it is designed to traverse. Indeed, it has been suggested that any ship having a maximum velocity less than a hundredth of the speed of light (1,860 miles per second) does not come into the category of a starship. (In this respect, of course, we are thinking *only* in terms of conventional space. The position regarding ships capable of swift or instantaneous transit through other possible dimensions is quite another matter.)

Starships having velocities not less than one-hundredth that of light traversing conventional space would involve voyages of the order of tens or hundreds of years. At present there appear to be only three possible types of propulsion technique:

1. Nuclear fission
2. Thermonuclear fusion
3. Matter/antimatter mutual annihilation

The basics of the first two are understood. The third is presently a more dubious concept, since it is uncertain whether or not antimatter exists.

It is essential that the energy from either of the first two sources be converted into a controllable propulsive force. An atomic and a hydrogen bomb are examples of (1) and (2) respectively, but neither can be said to represent *controlled* energy. On the contrary, once such energy is released it is most horribly and totally *uncontrolled!* Two basic methods for the conversion of such raw and awesome power into a controllable propulsion force have been envisaged:

1. Transfer of the energy released to a nonreacting propellant
2. Emission of reaction products

It is hardly necessary to comment that to realize either would not be without its difficulties and complications. A nuclear pulse rocket has been suggested in which deuterium-helium fuel pellets are ignited by an electron beam (deuterium is an isotope of hydrogen). Another suggestion envisages the detonation of small nuclear charges some distance *behind* the ship, though such a crude technique would only permit a fraction of the available energy to be utilized for purposes of propulsion—and it is the latter that is, after all, the object of the exercise. A more likely and practical possibility is the interstellar ramjet, so named because it would use interstellar hydrogen (there is quite a lot of it around) as a propellant and perhaps also as a fuel.

We cannot, and indeed should not, go too closely into this aspect for two reasons, both of them very valid. In the first instance, our technology is not yet sufficiently advanced to plan

a viable interstellar drive. Secondly, we cannot tell, even envisage, the sophisticated forms of interstellar drives that our
galactic superiors could, and probably already have, produced.

Let us now consider a spacecraft that falls into a category
somewhere between that of true starship and conventional
spaceship. Such a craft will obviously be something of a hybrid
and may seem of little real worth—too slow to have a genuine
interstar capability, too complex (and therefore too expensive)
for interplanetary use within the Solar System. At this point,
however, we should pause and reflect. Such a ship is indeed too
slow to have interstar potential *as far as we are concerned* with our
relatively brief life span of three score years and ten, give or take
a few. But should such a life span be accepted as the rule for
other worlds, as the universal norm? It is not even the norm here
on Earth. Dogs and cats have life spans of around a dozen years,
certain other creatures between a hundred and two hundred
years. But these are animals, not human beings, it might be
argued. Such a distinction is not valid biologically. We are all
living, oxygen-breathing, carbon/oxygen/nitrogen-based beings.
It is perhaps feasible that life spans greatly in excess of our own
occur among certain galactic peoples.

In the light of these possibilities we should perhaps look again
at our definition of a starship. Earlier, we defined such a craft as
one having a maximum velocity that exceeded one-hundredth
that of light (1,860 miles per second). This defintion is valid and
probably entirely acceptable to peoples like ourselves having
comparable life spans. But we must also accept as a starship a
spacecraft having a velocity less than 1,860 miles per second
(though much greater than that of contemporary, terrestrial
spacecraft) operated by galactic races whose natural life spans
are considerably in excess of our own. It may be convenient if at
this point we designate the two types as High-c and Low-c starships respectively (c being the accepted symbol for light velocity).

It is reasonable, we would suppose, to assume that several
Low-c starships could be constructed for the cost of just one
High-c vessel. (There is a contemporary analogy here, in that
several subsonic wide-bodied jets can be built for the cost of one

SST or supersonic transport.) In time, consequently, (suitable life spans permitting) there should be a greater profusion or density of Low-c vessels in space. The Low-c variety would require an inbuilt capacity to survive quite literally for a millennium or more in space. Another mandatory requirement would be the ability to store the propulsion energy for voyages of this order. The latter essential is probably rather less demanding than it seems, since a starship on a journey involving several hundreds of years is coasting for much of the time. Engine power is required only for acceleration, deceleration, and course changes—apart from relatively minimal amounts demanded by services within the ship.

It is considered in some quarters that the technology for Low-c starships is already within our grasp. This is a very debatable point, but assuredly if we do not have it now we should in the not too remote future. But with our life spans it could not be of much practical use to us.

Low-c starships therefore *have* the capacity to make long time-consuming voyages among the stars. In making a realistic appraisal of starship distribution in our galaxy, we should accept the fact that it may be this type that predominates. On a strictly economic basis the hypothesis is valid. Today it is cheaper by far to construct and operate a relatively slow commercial bulk carrier than a large, very fast warship.

It has been suggested that even at this moment radiations and particle fluxes generated within the propulsion units of alien starships are reaching our world. If this is so, location of the vessels responsible is not an automatic sequel. Galactic "noise" is at all times very considerable, that from "dying" atoms within the propulsion systems of sophisticated alien starships relatively puny, especially over vast distances. It is probable that this single fact constitutes the major problem in long-range detection of approaching starships—how to identify starship emanations from natural cosmic background. There are, however, a number of factors that might render differentiation possible:

1. Astronomical bodies do not possess relativistic velocities, i.e., though they all have a proper motion, their velocities do not

tend toward that of light (or even high submultiples of it) as might reasonably be expected in the case of starships.

2. Astronomical bodies do not normally exhibit significant changes in their proper motions, i.e, they do not accelerate and decelerate as we would expect starships to do. (An exception might apply in the case of stars that are members of binary or multiple star systems.)

3. The propulsion systems of starships would probably be modulated or pulsed, i.e, energy released not in a steady stream but as a series of "bursts." (A homely parallel here is provided by the automobile engine that gives pulses of power as gasoline vapor/air mixture is sparked and exploded in each cylinder.) Starship engine pulses would have a specific distinguishing frequency.

4. Astronomical bodies (stars and galaxies) emit considerable heat, i.e., they are exothermal, which is an identifiable characteristic. Starships, though undoubtedly emitting a measure of heat from their propulsion systems would, nevertheless, appear nonthermal. This may seem somewhat paradoxical, but it is associated with the fact that such vessels must not be permitted to heat up to the equivalent (black body) temperatures of their propulsion system output.

5. An alien starship that has approached to within a short distance of the Solar System would probably appear only as a *point* source of energy unlike astronomical bodies, which are, of course, of very large dimensions. This particular facet would depend to some extent on the dimensions of the starship and its actual distance from the Solar System.

6. Starships should possess a significant trajectory, i.e., they should have a point of departure and a destination, both being represented by observable stars. In our context this means we should, in theory, be able to identify starships leaving their home systems on a mission, an apparently very useful feature. Because of distance involved this is totally beyond our technological expertise at the present time. We must also remember that starships leaving a system ten light-years distant bent on an invasion of our world would be ten

years out on their way toward us when we observed their departure.

It is considered that most starships will possess at least some of these identifying characteristics and that one or more could conceivably become apparent during propulsive maneuvers in the vicinity of the Solar System, i.e., about one light-year out.

Let us now look at another aspect of starship detection. A useful parameter, were it obtainable, would be an estimate of the number of starships presently operational within the galaxy and their distribution. This is not possible because of our total (and understandable) lack of knowledge regarding the number of sufficiently advanced societies within it. About all that can safely be said is that the number must lie somewhere between zero and the maximum, whatever that may be. And the most we can say regarding the distribution of alien starships is that none appear to be making a "planetfall" here at the present time. Whether or not past visits have been made—surreptitiously in recorded times or brazenly and openly in geological times— remains a matter for conjecture and debate. Carl Sagan postulates ten thousand starships continually patrolling the galaxy with an average time between visits to any particular planet of a thousand years. It is difficult to comment on such an estimate but it is as likely to be as valid as any other. Right now it can neither be proven nor disproven.

An alternative way of looking at things is presented by G. V. Foster, who points out that since the motions of the stars within our galaxy are largely random, there occur from time to time opportunities for interstellar missions over distances considerably shorter than would at first appear. There is a certain truth in this, but the facts must not be taken too literally. At present, Barnard's Star, which has a very swift proper motion, is passing close to a star we call the Sun. Indeed, at the moment it is one of our closest stellar neighbors. There was a time in the past when it was not, just as there will come a time in the future when it will again be more distant. But the short distance between Barnard's Star and our Sun at the present time is short only in the interstellar sense. It is still a great, dark abyss of six light-years, roughly 36 million, million miles. It is also essential to realize that though "close" passages between one star and another are not over in a few days (or even years), neither do they happen overnight. Martians invading Earth (and once we believed this possible) would only have had to wait for a few years until a suitable opposition brought their world to within a mere 35 million miles of our own. A star people with the same idea in mind might have to wait for several thousands of years, rather a long time to keep invasion machinery ticking over!

Foster's point is well made, however, in that during the remote past beings from the worlds of another star could have made a brief "planetfall" here at a point in galactic history when their sun and ours were relatively close. Their technology was at that time capable of inter-star journeys of a few light-years.

The lack of observation of any starships within the galaxy at large (as opposed to any approaching the sun) has tempted some scientists to comment on what they consider an anomaly, a situation strangely inconsistent with the anticipated development and expansion of alien technologies. To the writer this viewpoint is inexplicable. Our galaxy, like all other galaxies, is vast, almost beyond human comprehension. It is some 13,040 light-years in diameter and contains approximately 100,000 million stars. In the present state of our technology we cannot expect to maintain total surveillance over such a vast cosmic entity. Indeed, it is at best doubtful if total surveillance could presently be maintained over the small "local" cluster of stars of which our Sun is one. If alien civilizations exist at various points within our galaxy (and it is surely more rational to assume that they do than don't), we can be reasonably certain that those civilizations millennia ahead of our own will be roaming the galaxy with an ease that should perhaps lead to our unease. Two battle fleets loose in the Pacific Ocean, were they not equipped with far-ranging aircraft and electronic eyes and ears, might well roam the great blue gray wastes for months on end, perhaps even years, without ever being aware of each other's presence. Starships at large, deep within the dark interstellar recesses of the galaxy, represent a monstrous extrapolation of this situation.

Our brief, however, is not the detection of alien starships at tremendous galactic distances. It is the potential, if any, for the detection of such craft approaching our Sun and its family of worlds. Undoubtedly the successful detection and identification of interstellar spacecraft anywhere within the galaxy would amply demonstrate the existence, not just of other races, but of highly advanced and sophisticated technologies. If, however, the spread of such technologies is reasonably uniform throughout the galaxy, successful detection techniques embracing our small corner would equally well enable this to be established.

On Earth, technologically inferior societies have, on occasion, become aware of those more advanced simply by sighting craft belonging to the latter. In the case of lonely Pacific islanders centuries ago, it was the sight of sails on the skyline. Today, primitive tribes deep within the great Amazonian rain forests become aware of other societies and their marvels by the sight and sound of high-flying jets. Carl Sagan sums up the situation well. "Primitive societies," he states, "could fail to observe the vast international radio and cable traffic passing over, around and through them but they could hardly fail to notice the most sophisticated product of this century in the sky above them."

In the first instance let us think about actual visual observation of a starship. This could be achieved either by virtue of external illumination or self-illumination or, alternatively, by occultation (i.e., partial or total eclipse of stars by the passage of a starship's opaque body across the direct line between star and observer).

External illumination of a starship can only be achieved by reason of starlight falling upon it. Stars are suns and are generally very liberal indeed in the amount of heat and light energy they radiate. In fact, most are quite prodigious in this respect. The extent of such light emanation, however, must be seen in the context of interstellar separation. We are all aware that a vast sun—perhaps many times larger than our own—is reduced by distance to a mere point of light in our night sky. Two examples well worth quoting are Deneb in Cygnus and Betelgeuse in Orion. The former is a celestial beacon several hundred times brighter than our Sun, and Betelgeuse is a red giant star of such mammoth proportions that were it at the center of the Solar System its vast bloated mass would reach out to, and beyond, the orbit of Mars. Despite these facts, both are merely bright little points of light in our skies. Since there are so many stars, surely there should be a cumulative effect. There is. We call it starlight, but the effect is hardly one of searing brilliance. This truth is made manifest any dark, clear, and moonless evening. The sky may be ablaze with stars but the amount of light reaching us minimal. Even on occasions when the ground is snow-covered and therefore highly reflective, starlight affords little in the way of illumination.

A starship in deep interstellar space is therefore unlikely to be brilliantly illuminated by reflected starlight, even allowing for the fact that in space stars appear brighter because of the absence of our light-absorbing atmosphere. Suppose, however, that a starship is in the final stages of its journey to the Solar System. Surely the fact that it is drawing very close to one specific star (in this case our Sun) means that a steadily increasing amount of light is impinging upon it. This is perfectly true, although a measure of caution must be observed. Such a ship on reaching the orbit of the outer planet Pluto is, relatively speaking, now quite close to the Sun. Pluto, however, lies at a distance of 5½ light-hours from the Sun and, although very large compared to a starship, is a very faint object in the skies over Earth. In fact, the reflected sunlight reaching us from Pluto is extremely slight, so slight that at magnitude 15 it is only discernible in quite large telescopes. In the circumstances, we must anticipate that a starship at a comparable distance from us, unless the material of its construction gave it a very high albedo (reflecting power), would certainly not be discernible. The vessel would have to approach much closer to the Sun and Earth ere we could hope to detect its presence merely by reason of reflected sunlight. From a defense point of view the warning time might therefore be unacceptably brief.

So far as self-illumination is concerned, this can only mean the glow coming from the vessel's propulsion system or associated cooling fins. Once peak velocity of star flight has been attained in the frictionless medium of space, the propulsion system is closed down. There is consequently no source of self-illumination in deep space, save on those occasions when the propulsion units are briefly switched on to effect course changes or corrections. At such a range, visual detection by reason of self-illumination can be safely ruled out. On close approach to destination planet, however, and perhaps for a period before-hand, we could envisage the use of retro-propulsion (reverse thrust) for braking purposes. Ships identified at this stage in the proceedings would, were they part of a hostile force, be much too close for comfort.

Let us see now if occultation would assist in the matter of starship detection. Any technique based on occultation must have two essential requirements satisfied.

1. The starship must be opaque to normal visible radiation.
2. It must also be sufficiently large (0.1–10 kms has been suggested).

The first requirement is patently obvious so that there is no difficulty in satisfying it. There is, of course, always the remote possibility that a starship of an ultrasophisticated technology might have been rendered transparent or translucent to visible radiation by judicious control of refractive index. But this seems so sci-fi in context, and in concept, that it can in all probability be safely ignored.

The second requirement is also straightforward. If a starship is to reveal its presence at a distance by eclipsing stars, it must be of fairly considerable dimensions. If it is sufficiently close to Earth, it will probably do this even though it is relatively small, but for any starship to do so on the outer fringes of the Solar System or beyond it must be of truly gargantuan proportions.

During the years immediately preceding World War II (1938 and 1939) it was quite seriously suggested by several people in England that German bombers in the obviously approaching conflict, might give away their positions and course to the ground defenses by reason of brief eclipsing of the brighter stars.

It was hardly a very realistic idea. For a start, the skies over the British Isles during the dark nights of winter are notoriously starless, because of clouds. Additionally, few persons had sufficient intimate knowledge of the night sky, its stars, and constellations to appreciate the significance of any such brief occultations, even if they were observable. The idea never got off the ground. When the German air raiders did finally come in 1940 they must have reached some prearrangement with the celestial powers, for never did they appear to eclipse a solitary star. Their understandable predilection for nights when the Moon was full or nearly so (the dreaded "Bomber's Moon") may have had something to do with this. The Moon must have been included in the arrangement, for they didn't even eclipse that. Perhaps people were too busy keeping their heads down to notice.

When we come to consider the probable dimensions of alien star vessels, we immediately find ourselves confronted by difficulty. Most likely ships, especially those of a task force destined to attack and take over our planet, would have to be of fairly considerable dimensions, if only to contain the vast numbers of "men" and amounts of material that the mounting of such an operation would presumably necessitate. This could, of course, be an entirely false premise. A few smaller ships equipped with ultrasophisticated weaponry might be entirely adequate to subdue Earth. The really large ships containing the alien colonists could follow. From a defense point of view, therefore, it is probably the earlier, smaller ships we would first need to detect.

The term *small* and *large* used in this context could be confusing, being only relative terms. So-called *small* starships could be quite large, rather magnificent creations from our point of view, and small only in comparison to yet larger editions of the same. In fact, the matter of starship dimension is a difficult one to quantify. Vessels designed to cross the interstellar gulfs by conventional methods will be quite colossal, but whether their size will be sufficient to give an occulting effect against the stellar background is a very moot point. Starships using extradimensional travel techniques (assuming these are possible) might be the really small fry. Since these would to us apparently materialize out of thin air (or should it be "thin space"), we wouldn't be concerned with any occulting effect.

An occultation technique of starship detection, were it to have any hope of success, would entail continuous monitoring of stellar brightness. For an object such as a starship to be detected, a minimum of three occultations would have to be observed. An example from the literature postulates an object having a diameter of one kilometer observable under certain specific conditions. For three occultations, the maximum object–observer distance works out to 5.2×10^7 kilometers. This is approximately equal to 28 million miles, which is barely the distance between us and the planet Venus at its nearest. Alien spacecraft approaching Earth at very high velocities would be on us soon after detection. It is felt that greater distances could be achieved, but with correspondingly lower chances of detection.

The task of monitoring the entire sky (and from a defense aspect probably nothing less would suffice) down to low stellar magnitude would be a daunting one. Since it is also one that, in the event, could so easily prove inadequate, it seems improbable that an occultation technique would ever be seriously contemplated.

To most readers, detection of alien starships by reflection or occultation will seem a very poor alternative to the use of radar. A few paragraphs back we pointed out that during World War II enemy aircraft at night could not be detected by star occultation. We do know, however, that the approach of hostile machines could be detected by radar long before they reached their objectives. Surely, here is the practical answer to the problem of hostile starships approaching our planet or lurking suspiciously just beyond the Solar System. Such a belief is further borne out by the fact that the detection and tracking of Earth-orbiting space vehicles and satellites by radar is already a well-established technique. Unfortunately, such methods are appropriate only for ranges up to a few thousand kilometers in the case of small objects like artificial satellites. The maximum achieved in this respect so far is probably the detection in 1968 of the "Earth-grazing" asteroid Icarus, which was located when at a distance of some four million miles from Earth. This distance corresponds to almost sixteen times that between Earth and the Moon. With larger objects such as starships this figure could no doubt be improved somewhat, though not to any considerable extent.

Very large increases in radar power would be required to detect spacecraft of small cross section over distances of interplanetary order. It has been estimated that to detect a starship having a cross-sectional area of 100 square meters at a distance of 0.1 astronomical units (approximately 9,300,000 miles) a power of 7.5×10^{12} watts per cycle would be required (assuming certain other parameters with respect to antenna gain and receiver characteristics were fulfilled). Since the frequencies used in radar are of the order of 1,000 megacycles per second, a power of some 7.5×10^{21} megawatts of power would be required to detect a starship lying at only a tenth of the distance between ourselves and the Sun (or in terms of the night sky, about seventeen times as far away as the Moon). As an early warning system, this has clear and definite limitations. We are assuming, of course, that our radar would be operating from Earth. Remotely operated radar probes sent by us to the very borders of the Solar System (near the orbits of Neptune and Pluto) are obviously a much better bet. By the use of such devices, we could in effect peer out into the darkness of the great interstellar ocean lying beyond the orbit of Pluto. Positive radar pulses indicating the presence of an object or objects could be relayed back to Earth in about five to six hours. Admittedly, we do not as yet have the technological expertise to set up such systems, but in view of the fact that we have already sent instrumented probes as far as Jupiter and Saturn, such prowess is not beyond us in the foreseeable future. Here then is probably the best bet with respect to starship detection—a number of remote radar sensing stations in different orbits. Some would orbit in the plane of the Solar System as already described, others in orbits inclined at a variety of angles to the plane so as to ensure detection of approaching visitors irrespective of direction.

Another technique envisaged is based on detection of the propulsion systems of starships. Clearly this can only apply if the systems are operating. The precise nature of the technique would depend on the form of propulsion being used. We would therefore have to be able to identify more than one system.

Detection of the propulsion systems of starships really involves detection of the resultant fluxes, i.e., the flow of propulsion products. It should be stressed therefore that this technique could only indicate that one or more (starships) were in the vicinity. Vicinity in this context means that region of the galaxy in which the Solar System lies. This is not the sort of technique that could be used to pinpoint the precise location and direction of a starship. It could, however, serve as a warning to initiate search procedures as soon as possible. A useful analogy might be the simple domestic gas leak. Our noses tell us the stuff is escaping. That does not clearly indicate the source of the leak, but at least we know to start searching; the warning has been given.

One form of starship propulsion envisaged is the use of deuterium—He^3 fuel pellets. These are heated to thermonuclear temperatures by an incident relativistic electron beam in which about 40 percent of the fuel is consumed. This leads to a very "clean" energy output, though neutrons result from a side reaction. These neutrons, it is reckoned, might be detected by appropriate apparatus 60 million kilometers (about 37 million miles) away. This is a distance only slightly greater than that separating Mars from Earth at a favorable opposition. It is hardly a vast distance by cosmic standards and all too short if the alien vessels were closing in on our planet at high velocity.

Another system mentioned earlier involving thermonuclear pulses of 10 kilotons' yield exploding some distance behind the starship would produce neutrons with a range of almost 800 million kilometers (about 500 million miles). This increases the previous range almost fourteenfold.

Yet another possibility in the propulsion field is the mutual annihilation of matter and antimatter. As mentioned earlier there is, as yet, no real evidence that antimatter exists. From such a system the flux would have an estimated maximum range of 800,000 kilometers (about 500,000 miles), a relatively short distance.

There *are* thus possibilities for detecting the presence of certain types of starships but only at distances of an *interplanetary* order. It should be stressed that the figures quoted with respect to flux range are optimum ones. In practice, they would probably be lower. A starship, if it is going to be a practical and efficient craft, must minimize energy dissipation that is not in the form of propellant kinetic energy. *Our* figures relate to hypothetical and, we must also remark, somwhat inefficient starships.

At the moment it has to be admitted that the chances of our civilization here on Earth becoming aware of the approach of alien spacecraft are not overhigh. So far as early warning goes, we are presently wide open to a cosmic "Pearl Harbor."

Round-the-clock saturation searching by radar sensing currently offers the most viable answer. At present, this has a very limited scope, but the possibilities will, no doubt, increase as our technology advances. If, in the end, an outer defense system of solar orbiting radar stations can be set up as described earlier, these ought at least to give some warning of an approach. But it must be remembered that electronic warfare is something in which both sides can engage. During World War II the Allies dropped strips of metal foil from aircraft to confuse the Germans. It was a crude, though fairly successful, ruse. Advanced beings coming at us from the depths of interstellar space might easily have a technique capable of rendering our most sophisticated radar systems totally inoperative.

A recent rather interesting suggestion might form the basis for a second or inner line of defense. This envisages a spherical array of 1,000 spacecraft at a distance of 2 astronomical units from the Sun (186 million miles), each having an effective range of 16 million miles. By this means it is hoped that any intruder passing through the "shell" thus formed would be detected. A project of this extent represents a massive undertaking, and it is difficult to conceive of our terrestrial civilization embarking upon one unless it could clearly be shown that our planet was in imminent danger of attack.

Radar and other forms of electronic sensing are almost certainly the long-term answer to detection. It is to be fervently hoped that if attack by starships ever threatens, technology will have had adequate time to develop the techniques required. In this form of attack there would almost certainly be no second chance for us.

5. BASIC DEFENSE

In the preceding chapter we made a fairly detailed study of the "early warning" aspect. The conclusions reached perhaps inevitably were a little less than optimistic—at least so far as the present and immediately foreseeable future are concerned. We must hope that if a threat from deep space should ever occur, terrestrial technology will, by that time, have progressed sufficiently to supply some kind of answer. The premise would seem more reasonable if we could be sure that by then the nations of Earth had stopped their inane and senseless bickerings. Those who play around in the gutter rarely see the stars!

Mere warning of an impending extraterrestrial attack does not, of course, in itself constitute either a defense or a deterrent. All it does is provide time, an absolutely essential breathing space, which must be used wisely and judiciously. In the circumstances we are envisaging in these pages, we would require also the weapons and the techniques to combat effectively a cosmic foe who could easily be ruthless. Without these things, a warning, however long, could not alone help us. We do not in strict military terms have to defeat the enemy from the skies—only to prevent his establishing a foothold. If we make it too difficult for him, too costly in terms of "men" and materiel, if we create a stalemate, he could decide to transfer his aggressive operations elsewhere.

In the pages of science fiction, the arrival of the cosmic invaders represents the great moment—the climactic episode when the combined mights of all the terrestrial space fleets sally forth to do battle with the invaders ere they reach Earth. In future centuries such defending fleets may well exist. For the present and immediately foreseeable future they do not. Consequently our defenses, assuming any are possible, must be much more basic, perhaps even primitive in character. Two questions immediately arise. Could any such technique possibly prevail against the offensive capabilities of a cosmic host having a technology perhaps a thousand years in advance of our own? And could a quick, cost-effective defense be adequate in holding or deterring the marauders at a safe distance from the surface of our planet? To some extent this seems like pitting bows and arrows against modern automatic weapons.

One inexpensive and seemingly viable possibility was suggested in a science fiction short story dating back to the summer of 1939. In this particular instance a very large alien spacecraft (presumably a starship from interstellar regions) had reached the Solar System and had apparently gone into parking orbit above the Earth. Since the story was written before the advent of either radar or space satellites, the approach of this great alien man-of-war had gone unnoticed. The occupants of this mysterious intruder, being representatives of a highly advanced, but not highly humane, cosmic society, had presumably come to the conclusion that representative samples of Earth's surface, its flora, fauna, and artifacts should be scooped up here and there by means of a gigantic mechanical shovel arrangement. This highly improbable but distinctly lethal device descends scythe-like from an apparently innocuous sky to gouge a great furrow on Earth's surface, sweeping the contents of its mammoth grab back up to the ship for detailed examination. The idea seems truly bizarre, but at least a high mark is due to the author for originality. It is doubtful if any one, even then, was really prepared to take such an idea seriously. Clearly, the mind of the author had been influenced by two concepts—the deep sea trawl and the steam shovel. Combined with alien beings they made a most unusual triumvirate. The cover artist portrayed the grab at the end of a long arm (which disappeared into the clouds) ploughing its path through the very heart of some unfortunate metropolis. In its great maw was everything from human beings to streetcars.

It is not this unusual device that interests us, however, but the means of defense that was devised against it. At the time the story was written terrestrial society had, of course, no space potential whatsoever. Consequently, a means of combating the alien vessel and its frightful sampling device had to be one that could operate from the surface of the Earth. Without space vehicles and having only piston-engined aircraft of very limited ceiling, the "thing" was effectively out of reach so far as mankind was concerned.

Following more disastrous and sanguinary grabs by the cosmic shovel, it was decided that the best and probably the only means of effective defense (and retribution) lay in the manufacture and launching of innumerable balloons, each bearing a sensitive, high-explosive charge. What, in fact, was envisaged was nothing less than the hasty creation of an immense upper atmosphere minefield. And so, night and day, without cessation, hundreds of plants throughout the world commence mass production and launching of the deadly balloons. Production schedules begin to falter after a few days, however, as the "grab" continues its lethal work, apparently unimpeded and unharmed by Earth's crude attempt at a defense. Worldwide panic and demoralization is already setting in when, neatly, almost on cue one might say, a massive explosion of unprecedented violence and brilliance takes place high above the earth. Minutes later the shattered flaming remains of the great starship are seen hurtling downwards through strangely roseate clouds—all great rousing stuff, however improbable. It seems most unlikely that a vast starship would descend into the upper layers of our atmosphere and even less so that a relatively small chemical explosive charge—or even a number of them—could result in its total destruction. But, wonderful things like that can happen in science fiction, which no doubt is part of its great and growing appeal.

Is there any kind of validity about the aerial minefield as a defense against possible cosmic marauders of the future? The laying of a minefield at sea is an old and still fairly effective method of protecting harbors, ports, and estuaries from enemy attack and intrusion—from the sea at any rate. And the clandestine sowing of mines in the sea lanes during the two world wars brought about the destruction of many vessels, even great dreadnought battleships. Is then our analogous technique possible when it comes to warding off cosmic predators?

In the story just described, the mines were apparently simple contact mechanisms borne aloft into the upper atmosphere by balloons filled with either hydrogen or helium. Such devices would certainly be inexpensive and easy both to construct and launch, but to create a minefield of sufficient extent to cocoon the planet would be a well-nigh impossible undertaking. A starship in parking orbit would be well above the devices, though they might effectively combat or seriously restrict the freedom of small attack and shuttle craft emanating from a mother ship.

What we really require, however, is some form of protective screen whereby some of the invaders could be destroyed and the rest dissuaded. In this context we are thinking primarily of a cosmic minefield spread quickly, effectively, and surreptitiously ahead of advancing hostile alien spacecraft at some distance from our planet.

Thermonuclear minefields laid in deep space and capable of detonation by the mere close proximity of an alien marauder would appear to represent something of a defense potential. The most obvious difficulty is, of course, in deciding just where to place them, remembering that objects "placed" in space do not remain "placed"; they begin to orbit something, generally either a planet or, if far enough out, the sun itself, in the fashion of little planets.

A belt of thermonuclear charges with proximity fuses around the Earth seems at first glance to represent a reasonable form of defense. It would have to be, however, a concentric shell around our planet to prove effective. And even then, sensing or repelling devices on the alien ships could allow them to pass through unscathed. Indeed, the defenses might prove more lethal to terrestrial spacecraft trying to pass through in peace!

A perhaps more feasible and realistic proposition would be for Earth's defenders to construct and launch a number of orbiting minefields, each individual mine containing a power unit capable of taking it out of orbit in response to a transmitted signal. This sweeping artificial "comet" of deadly charges could then be streamed across the path of the advancing space enemy. This again assumes that the invaders are not equipped with repulsion fields by means of which the mines could be swept aside. It also assumes that we have a warning and a tracking system enabling us to move the minefields on to their proper course in time.

At the present point in human affairs a scheme of such magnitude and complexity is still beyond our capacity to achieve, though it might become feasible in the not too remote future. Certainly it represents a defense line of sorts. Whether it is one that could be held is another matter. It could, quite easily perhaps, be swamped by sheer saturation tactics. This assumes that the aliens propose to carry out their operations using several vessels. To defeat, occupy, and subdue our world would probably render several vessels essential. "Cometary" minefields of the type we have been envisaging could be activated either from Earth, the Moon or space stations orbiting either. Indeed, with a view to interception farther out, the lethal "streams" could be placed in parking orbit around Mars or around the major satellites of Jupiter.

Perhaps an even more practical adaptation and extension of the same basic concept would be to have patrolling space mine-layers—craft able to move quickly into the invaders' paths and there strew their highly lethal cargoes. Such operations carried out at a distance from Earth, perhaps well beyond the orbit of Mars, might prove especially advantageous. At this point the velocity of the incoming craft could be so great as to render evading action difficult. But, and this must be emphasized, the Solar System, though incredibly minute by galactic standards, is still very vast indeed compared to ourselves. Thus, a considerable number of these vast, maneuverable space minelayers on continual patrol would be essential. Such a scheme would no doubt be very costly. Governments do not, on the whole, like high cost defense systems unless the need for them can be very plainly seen. An impending invasion by beings from an alien solar system would be justification enough and more. Unfortunately, the time interval between realization of the awful thing that was afoot and the creation and deployment of such a force would almost certainly prove inadequate.

It cannot truly be said that essentially passive defense systems of the sort we have been considering could do other than delay a rather terrible march of events. To ensure the destruction of an invasion attempt virtually *all* the invading starships must be either destroyed or compelled to turn back. The arrival of just one, considering the type of lethal and sophisticated weaponry it might carry, could all too easily be one too many so far as our planet and its peoples were concerned. Moreover, the total rout of one wave could be followed by the arrival of a second, third— and a fourth. It could so easily prove to be a "no-win" situation—for us.

6. BEWARE THE "WHITE HOLE"!

So far we have considered detection of alien starships approaching Earth and the possibility of a basic defense technique. In so doing we have glossed over one rather important, indeed highly pertinent aspect—that of transit. If our world is to be reached by some of our neighbors in the galaxy, with hostile intent or otherwise, they must have the ability to cross the tremendous abyss that separates star from star.

In the forerunner to this book *(Interstellar Travel: Past, Present, and Future)* I dealt at some length with a number of techniques that might enable advanced cosmic communities (and one day perhaps ourselves) to flit quickly and easily from star to star or even to circumnavigate the galaxy. These embodied the concepts of hyperoptical velocity, cutting through curved space, and the use of black holes as convenient "space tunnels." It is not, therefore, the intention to dwell further upon these matters in these pages. When dealing with the potential of black holes in this respect, however, a very brief reference was made to the possibility of a suprasophisticated starship, the product of an equally suprasophisticated technology, being able to create its own hyper-gravitational field at will—or, in a sense, to create its own personal black hole "space tunnel." It might now be a sound idea to expand on this theme, for two reasons. In the first instance, the concept does seem remotely feasible, although, of course, far beyond our capabilities now and for the foreseeable future. In the second, such a device could well be used as a weapon of the most devastating and frightful kind. Quite recently it was suggested that the tremendous and somewhat inexplicable explosion that occurred in the Tunguska region of Siberia on June 2, 1908, was due to a collision between Earth and a black hole of fairly small proportions. It is doubtful if this theory is really viable, but it is certainly a novel and interesting one. If there exists even the remotest chance that the Tunguska event can be attributed to a small black hole, then the recurrence of such an event is to be greatly and rightly feared, for it devastated nearly 30 square miles of forest and tundra, vaporized herds of reindeer, melted metals miles from the impact center, and sent violent atmospheric and seismic shock waves around the world. Had a city lain in that particular area, both it and its inhabitants would have been "taken out" in the fullest sense. In fact the city of Saint Petersburg, now Leningrad, had a narrow escape, for had the "thing," whatever it was, impacted a short time later, the earth's rotation would have brought Saint Petersburg into the target zone.

Figure 5A
Appearance of a star that has become a Black Hole

A black hole is a region in space where matter vanishes from this part of the universe and emerges through a white hole in another—or so, at least, we presently postulate. Basically, a black hole is caused by the total collapse of a star and the consequent creation of a region where gravitational force becomes virtually infinite. We are all more or less familiar with gravity and what it does. If we raise an object and then release it, it at once falls to

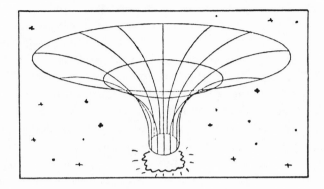

Figure 5B
Space near "collapsing" star curved by gravitational field of that star

the ground. If it failed to do so, we would all be very surprised. We would probably be more than a little troubled as well, for the force holding us to the ground would also have vanished, and presumably we'd be floating or hovering around the room. The force concerned is, of course, that of gravity, the fundamental force of the entire universe. It is keeping the writer of this book in his chair. It is also keeping him (and all his readers) on the surface of this planet. Additionally, it is entirely responsible for maintaining that planet in its eternal orbit around the Sun and the Sun (plus all the multitudinous other suns) in orbit within the galaxy we know as the Milky Way—a very fundamental and basic force indeed. To somehow harness and concentrate such a force would be a very great feat. As we shall presently see, the inherent possibilities could be most remarkable.

An interesting and novel example from the world of science fiction comes to mind at this point—once again from a short "pulp magazine" story of the 1930s. Though it is hardly possible to endorse the particular possibility the author envisaged, the general theme gives a unique idea of the nature and power of gravitational attraction.

The story opens in a way familiar to much science fiction of that period—a world war of the future in which vast aerial fleets devastate the cities of the opposing sides. Enter at this point the superscientist and his latest brain child—a sort of monster electromagnet that, instead of producing an orthodox magnetic flux affecting only ferrous metals, radiates intensely powerful *gravitational* waves capable of plucking anything out of the sky. Not surprisingly, this plays havoc with the raiding air fleets of the opposite camp. Unfortunately for the side possessing this powerful but improbable device, a swarm of large meteors is due to cross Earth's orbit at a point not far from Earth itself. Normally it would pass harmlessly by, but now the effect of the artificial gravity field is to pull the meteors toward Earth and the source of the field. In the ensuing holocaust the nation using the device is devastated in an appalling manner as huge meteorites change its surface to one resembling that of the Moon.

One has to confess that all this seems a highly improbable state of affairs. Perhaps one day such a device, or at least something along these lines might be produced, but the effect on meteors seems much too convenient. Meteor swarms are certainly not unknown and are believed to be the remains of former recurring comets. At set intervals during the year, especially in the fall, they can, and often do, give rise to spectacular showers of so-called "shooting stars." (The Leonids in November 1966 were a classic example.) It must be emphasized, however, that the objects concerned are rarely much larger than grains of sand. A swarm comprising pieces large enough to devastate much of a country is unlikely. But undoubtedly the author of this story gave a very graphic slant to the concept of gravity, even if he did permit himself a high degree of astronomical license.

Black holes represent the manifestation of gravity in its most extreme form and present us with the theoretical possibility of "wormholes" or "space tunnels" to other parts of the universe or, in other words, shortcuts. Unfortunately, black holes are unlikely to be sufficiently commonplace or conveniently placed to serve as interstellar detours. The question then arises of creating such space/time vortices artificially when and where required. If the thing is possible, an alien invasion force would probably elect to create its black hole tunnel entrance at a point near its own star system. We thus have the general idea of all the starships involved in the operation passing through a single common "tunnel." But in such an extrapolation of the laws of physics and the powers of technology, the effects might be even more bizarre than we think. We could find ourselves confronted not by invaders from another star but by marauders of strange form from *another galaxy!*

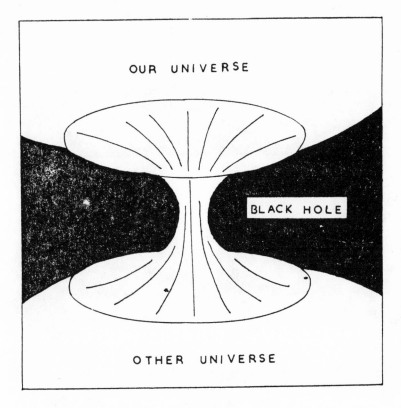

Figure 6

An Einstein-Rosen "bridge" or "wormhole." The upper portion links our universe with the event horizon; the lower carries on into another universe

The germs of an idea to achieve a "manufactured" black hole are already beginning to appear in certain scientific journals. Basically, the concept is one of starships sweeping space ahead of them with intense magnetic fields, allowing the particles of ionized interstellar hydrogen scooped up to enter the fuel tanks, thereby powering the fusion-engined craft. It is envisaged, however, that only a very small proportion of this material would be so employed. The bulk would serve the prime purpose—that of accumulating matter and bulldozing it forward. At a certain point in this process the accumulation of matter would lead to the creation of an artifical black hole. The space vehicles performing this sweeping and concentrating process would probably have to be regarded as expendable, since it is more than

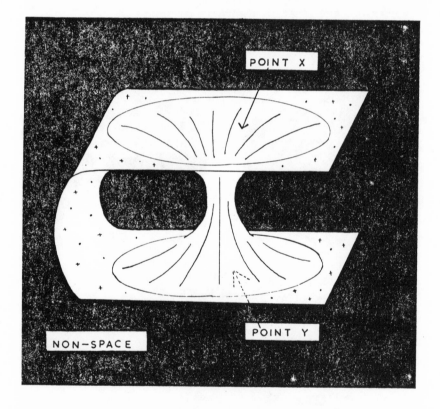

Figures 6A and 6B

Two other interpretations of the wormhole, conceivably connecting two well-removed points in our own universe

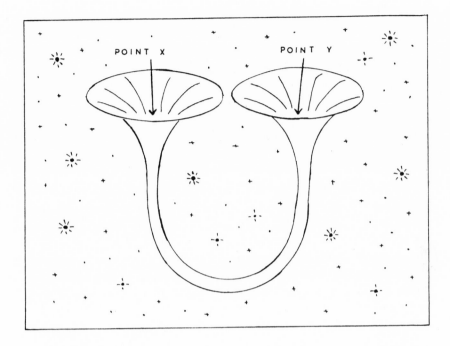

likely that they would themselves be sucked into the black hole as it began to form, there to be swiftly crushed out of recognition, indeed out of existence. Automatic unmanned ships would be called for, as presumably crews would prefer not to play a kamikaze role.

The major snag in this scheme is the amount of matter that the "scavenging" spacecraft would have to amass at the chosen point in space to lead to the creation of a black hole of sufficient dimensions. One having a mass equal to that of the Sun would, it is estimated, only possess a diameter of 3.7 miles, and a navigable aperture of a mere 198 feet: Rather a tight fit if a fleet of starships were to pass safely through, even if they proceeded single file. A black hole equivalent to five solar masses would have a diameter of 18.4 miles and a navigable aperture of 984 feet—still barely adequate. Unfortunately, to achieve anything very much larger brings increasing difficulty and added danger.

For example, a black hole having a mass equal to a million times that of the Sun would have a diameter of 3.7 million miles and a navigable channel of 37,000 miles. Such a vortex in the vicinity of a star system would represent a grave peril to the system because of the risk of disruption engendered by the intense gravitational field. A black hole equivalent to ten solar masses is considered a reasonable compromise, both on grounds of feasibility and safety. For the record, such a black hole would have a diameter of around 37 miles and a navigable aperature of about 2,000 feet.

It would, no doubt, be much more convenient to have starships each capable of creating its own little "tunnel" in space when and where it chose. This seems to us mere terrestrials at least a total impossibility. Black holes arise as a result of the inward collapse of vast quantities of matter causing a gravitational field of almost infinite intensity. We can conceive of no other physical parameter that would permit the creation of such a "wormhole" in the fabric of space/time. This, of course, might only be a measure of our intellectual inadequacies in this field. Therefore, though the idea of individual vessels each being able to form its own black hole does seem very unlikely, it would probably be unwise to rule out the possibility entirely. Such a power also has a direct bearing on the production and use of very small black holes as a weapon. Although this possibility might also seem mercifully remote, we cannot ignore it either. To primitive tribes in the rain forests of the Amazon basin, a thermonuclear bomb would, assuming they could envisage the physical process, seem equally, perhaps even more, improbable. It just might be possible for aliens to collect and process sufficient interstellar material in the environs of our Sun and cast this devastatingly in our direction in the form of minute black holes. Against such a form of attack it would be very difficult indeed to devise any form of defense. Even the highly improbable colored rays of the more lurid science fiction magazines seem a slightly happier prospect.

In discussing synthetic black holes as a means of interstellar transit we should really have emphasized that, if a black hole is to be used in this way, it is essential that it be of the *rotating* variety. Theoretically, in passing through a nonrotating black hole (assuming such a thing can exist), it would be impossible to avoid the singularity at its center. Ship and contents would be crushed into nothingness. Since all objects in the universe appear to rotate (from the smallest asteroid to the largest galaxy), it seems reasonably safe to assume that black holes do likewise. Whether or not this would apply in the case of those artificially produced remains an open question. If it did not, it would certainly have to be set in motion. We would assume, however, that a civilization with the technology to create black holes artificially would also be capable of giving them a spin.

Thus far, we have really only considered the *departure* of aliens toward our Sun and Solar System. If current theories are correct, their ships enter the black hole they have created and thereupon disappear from view. What, it must now be asked, are the circumstances of their arrival in the neighborhood of the Sun? In the first instance it would probably be a good plan to examine contemporary cosmological thinking on natural black holes.

There is one respect in which black holes are considered very unsymmetrical. Although matter can flow into them in the greatest profusion, none whatsoever can flow out. They could, in fact, represent one-way valves. So far as our galaxy is concerned, a recent estimate of matter consumption by black holes (for what it is worth) is one solar mass per day. Since the age of our galaxy is reckoned to be around 10,000 million years, it should, by the rules of simple arithmetic, have ceased to exist after only about 270 million years. Clearly it has not; therefore some kind of compensating mechanism or process must be at work.

In 1935, Einstein and Rosen produced a mathematical treatise that postulated matter disappearing at one point in the universe and reappearing in another. This is remarkable in a number of respects, not the least being that, at the time, the idea of black holes had not even been considered. Today, the belief is that production of a black hole at a point in the galaxy brings about the formation of *a white hole in another,* the two being joined by an extradimensional "space tunnel." Thus, matter flowing into the black hole emerges from the white hole, and it is this process that is seen as the compensating mechanism. The eminent British cosmologist, Sir James Jeans, predicted something along these lines even earlier when, in 1928, he wrote "the type of conjecture which presents itself, somewhat insistently, is that the centers of the nebulae are of the nature of 'singular points' at which matter is poured into the universe *from some other spatial dimension.*"

If the black hole tunnel/white hole theory is valid, we must suppose that a black hole created artificially for purposes of quick, extradimensional galactic transit would also lead to the equally synthetic evolution, somewhere else in the galaxy, of a white hole. Aliens using such a system to visit or to invade our world would disappear down their black hole "space tunnel" to emerge, perhaps almost instantly, from a white hole that had recently and mysteriously appeared just beyond the Solar System. The hint of a very effective defense immediately suggests itself—destroy the white hole and at once! What unfortunately does not immediately suggest itself is how this could be done

The question of white holes is most interesting and intriguing and seems to fit into the fabric of the theory very neatly. As Sir Fred Hoyle puts it, "The structure developed by a local collapsing object (i.e., a star) consists not just of a black hole, but of a black hole together with a white hole."

The location of black holes within our galaxy is not at all a straightforward operation. Quite the reverse, in fact, since to all intents and purposes black holes are, understandably, invisible. Neither light nor anything else escapes from a black hole and indeed the only means whereby the presence of a black hole can be detected is when the star responsible for its creation was originally part of a binary or double star system. In these circumstances the surviving (visible) star either emits violent outbursts of X rays or displays a wobble in the proper motion, or both.

White holes, on the other hand, should, it seems, be easily visible and therefore detectable with conventional astronomical telescopes. So far, however, it must be stated that none have been positively identified. This is, perhaps, not too surprising, since a white hole might be indistinguishable from a distant cloud of gas or some other celestial object. Chance could easily have decreed that white holes at present be formed in very remote parts of the galaxy. In this field these are early days yet and we must be patient.

Aliens using the transit technique of an artificially generated black hole or gravitational vortex would have to consider a very real and practical problem. Once they had entered the thing they would, like Caesar, have crossed the Rubicon, in this case a celestial one. Once having made their exit near the Solar System via a white hole they would have no way back save through normal space which, as we all realize, is a long and decidedly arduous business. Matter can enter a black hole but not leave it; matter can leave a white hole but not enter it. It is, or at least could be, as simple as that—for galactic travelers a rather nasty "impasse." The system constitutes a *one-way tunnel* in space. Our invaders would therefore have three choices—victory and dominion over us (their avowed aim), a stalemate allowing them to coexist in harmony with us (in the circumstances a most unlikely prospect), or defeat and extinction. It might be asked why, if they had to return, could they not proceed to the creation of a new black hole. We must stress again, however, that black hole creation must be seen as a vast and protracted piece of cosmic engineering. It is unlikely to be something that could possibly be achieved with limited resources in a few days or weeks. So at least *we* believe—but if their abilities do enable them to perform such seeming miracles, then presumably so much greater could be the danger to us.

So far we have cheerfully ignored another vital matter that potential space invaders would first have to solve. To create an artificial black hole would be only part of the problem. If, for example, they inhabited a plant orbiting Capella and were bent on reaching the environs of the Sun, they might be a trifle upset to emerge in the vicinity of Betelgeuse, a red giant star that ages ago swallowed up its family of planets (assuming of course it ever had one). Transit between stars via artificial black hole shortcuts is only valid if the position of the corresponding white

hole can be selected unerringly. As things stand at present it
appears much more likely that the white hole would appear at a
point in the galaxy dictated not by choice but by certain in-
alienable physical laws. The distance between black hole entrance and white hole exit might bear some mathematical
relationship to the mass of the two holes, but this so far is the
purest conjecture. That, in itself of course, if valid, takes care
only of distance and not of the equally important orientation
aspect. We have as yet no means of working these things out.
Aliens using the technique no doubt could.

Another difficulty from the invaders' point of view could be
the gap between the formed white hole exit and the Solar
System. So far we have merely assumed they would be near each
other. But "near" in this context could still be something of the
order of a light-year or a little less—still a lengthy journey even
for beings reaching us from the other side of the galaxy. This is
equally valid with respect to the generated black hole entrance,
which would clearly constitute an object of considerable danger
were it to form too close to their home star and system of
planets. So our potential invaders, though they might be enabled to make near instantaneous transit through the tunnel in
the space/time continuum, would, in all probability, still be
confronted with a normal space journey of between one and two
light-years. But perhaps photon drive combined with time di-
latation effect would solve the problem of "short" interstellar
hops like this.

There remains one other not inconsiderable problem, probably the most bizarre of all—the fact (belief would probably be a
better term) that transit through a natural black hole could also
be a journey into another time. In other words, the space tunnel
could also be a time tunnel. The theory behind this gem is, as
might be imagined, somewhat involved. It would appear though
to possess a certain logic. If there is anything in this, then, we
must inevitably wonder whether or not an artificially generated
black hole would possess the same potential. The prospect opens
up a number of amazing possible paradoxes. The invaders issu-
ing from a suddenly appearing white hole could be representa-

tives not just of another civilization but of one from another time. And if the tunnel were merely a time tunnel, perhaps even ruthless invaders from our dying Earth of the future. The mind begins to boggle!

Here too, we must assume that aliens using the technique are aware of all the possibilities and are able to avoid the peril of arriving in another time rather than in another place. It would be a bit upsetting for them after all if they had left a dying planet to wind up not beside a pleasant contemporary Earth but an Earth of the future as moribund or nearly so as the one they had left. At least it would be an intriguing form of poetic justice!

If the creation of black hole space tunnels is technically feasible, then presumably we could anticipate a number of the things to exist even now within our star-strewn galaxy, the fantastic product of highly advanced cosmic technologies. We would not like to go as far as to suggest that the galaxy is "honeycombed" with them like some glorified anthill, but it could be that the beloved science fiction ideas of galactic federations and star wars have a certain validity after all—and that is quite a thought! Colonization by even one star people could lead to the process being repeated again and again as the cosmic frontiers of that race are continually enlarged. Given long enough (and galaxies endure for a very long time) and assuming that a sufficient number of galactic communities are at this game, then sooner or later it would seem the Solar System and our planet *must* be in peril.

There could be a potential and rather special danger if black hole transit or any other extradimensional alternative (such as cutting through curved or folded space) is feasible, inasmuch as we could receive little or no warning of the impending attack. Earlier, we considered the possible detection of approaching starships. If a white hole were to open up very close to the Solar System (or even within it), we could be inundated, taken completely by surprise by the emerging hostile space fleet. And should the idea of cutting through normal curved space into hyperspace or nonspace and back again (as outlined in the pages of *Interstellar Travel*) be valid, the position might be even more menacing. In these circumstances, there would not even be the brief warning afforded by the preappearance of a white hole in the vicinity. The fabric of space/time close to our world could suddenly be rent asunder to allow the materialization of ships, aliens, and weapons beyond our wildest and worst imaginings. And that would be something we could do very little about.

7. "WATER, WATER EVERYWHERE..."

It is reasonable to suppose that an attack on our planet by an alien task force from a world or worlds of some other star would be with the intention of taking it over, either because of the suitable "lebensraum" it offered or the valuable war materials it contained. In either of these circumstances the existence of an indigenous population would in all probability represent a hindrance and a danger. Though much would obviously depend on the ethics and morals of the race concerned, one bent on aggression and aggrandizement on this scale might feel quite uninhibited in its attitude toward terrestrials. Throughout recorded history our own standards in this respect have hardly been of the highest. Indeed the present century has witnessed genocide on the greatest scale yet. In 1945 the world was sickened by the revelation of what had been done to the Jews in Europe during the preceding years of war. Much more recently we have seen with equal revulsion the treatment meted out by a fanatical ultra left-wing regime in Cambodia.

To remove or otherwise dispose of Earth's natural inhabitants migth thus seem a perfectly logical step to alien minds. The invaders could first take over the planet and then embark on a policy of liquidation. On the other hand they might elect to streamline the operation by divesting themselves of the indigenous population during the takeover process. Although the carnage that would undoubtedly ensue in a straightforward alien attack with ultrasophisticated weaponry would be tremendous, it is unlikely that it would account for all the troublesome terrestrials. The objective might be achieved by the use of bacterial warfare, but that could prove too slow and might also lead to further problems. In any case, bacterial warfare can easily prove two-edged. The aliens might, instead, consider a novel possibility—that of allowing some deliberately initiated natural catastrophe to do the deed for them. Several probabilities spring to mind, such as induced earthquakes and cataclysmic volcanic eruptions, both on a vast scale. No doubt if the almost legendary San Andreas fault could be activated in some way, a large part of California would be devastated; a triggered Mont Pelée somewhere might rend itself asunder, wiping out all the inhabitants of a large city just as the prototype Pelée did to the thriving city of Saint Pierre on the island of Martinique on Ascension Day, 1902. But artificially generated events even on this scale could hardly achieve the full Machiavellian alien purpose. There would seem to exist only one agency, only one means by which the entire population of this planet could be expunged quickly, effectively, and permanently—quickly raising the level of the great oceans and seas that cover the major portion of our world. This prospect might seem like the tale of Noah and the legendary biblical flood all over again—but on the really grand scale. Just how could the oceans be induced to serve the role of mass executioner on behalf of alien invaders?

The idea of a drowned world is certainly not a new one. The present writer's first contact with it dates from boyhood days when a serialised weekly story on this theme appeared in the pages of a boy's paper of the period. The opening installment, which dealt with the actual drowning of the Earth, was extremely graphic, and clear memories of it still linger. On this occasion the sudden and catastrophic rise in the level of the oceans was brought about by the close passage of a strange body from beyond the Solar System on which the name Viro had been bestowed. Warnings relating to the nature of the impending catastrophe had been widely circulated. These had told the peoples of Earth to construct boats and prepare for a life afloat. Unfortunately, virtually all had decided to construct their private arks at the water's edge and allow the rising water to float them off. The family whose future activities were to constitute the tale of life on an inundated world, however, was considerably more astute and far-seeing to the extent of constructing its ship on the summit of a mountain, despite the jibes and jeers of its fellows. When the flood raised by the close passage of Viro reached the area, it came not as a peaceful, slowly rising sea but as a tremendous roaring wave many hundreds of feet high. The effect was most eloquently described—the great grey green wall of water rushing inexorably landwards, the vessels constructed at the shore swamped and shattered in an instant, buildings and bridges swept away and, most poignant of all, the lights of cars rushing up the winding mountain road, their terrified occupants striving desperately to escape the terrible waters. None of them made it to the top—not that it would have made a great deal of difference, since the entire mountain was enveloped by the flood. As the waters reached and closed over the cars, their headlights turned green briefly before being doused forever. As might be expected, the vessel at the summit of the mountain floated safely away to be borne for months over a superocean covering the entire planet.

The passing years have obliterated the memory of all the perils and adventures that followed, with one solitary exception. On a certain day, the coordinates of latitude and longitude indicated that ship and occupants were sailing *over* New York City. Three of the crew in a small submarine sailed up the great undersea canyon that had once been Fifth Avenue. From the ship itself could be seen a peculiar vertical rodlike object projecting a few feet above the water—part of the radio mast atop the Empire State Building. As a point of interest, this modern day (1932) Noah's Ark had no pressing fuel problems, being powered by three slender funnel-like rotors which converted wind power directly to electrical power. Today no doubt, the author of this tale would have settled for atomic propulsion

The close passage to Earth of another planet would certainly raise immense tides that would inundate a very considerable portion of our world's land surface. The extent and height of such tides would, of course, depend on the size of the rogue planet and how close it came to our own. Once it had passed, the ocean levels would begin to subside, assuming the vast polar ice caps had not melted in the process. How though, could such an effect be artificially induced? Surely even the most technologically advanced aliens could hardly arrange for another planet to make a close pass to Earth.

Figure 7
How the raising of the ocean levels due to melting of the polar ice caps would affect the British Isles

There are three possibilities. Large scale inundation could probably be achieved if the invaders were to melt the polar ice caps—perhaps by the use of superlasers or massive thermonuclear devices. Alternatively, large meteorites or even small asteroids maneuvered from their natural orbits could, by plunging into the oceans, raise immense tidal waves that might inundate whole continents. The third possibility is the use of some form of powerful gravitational beam whereby the waters might be raised in a manner similar to that brought about by a passing planet.

Of these three stark possibilities, the first would probably prove the least effective, being neither total nor particularly fast-acting. For the record, however, recent estimates of what melting polar ice caps might achieve are impressive enough by any standards. Sea level, it is reckoned, would rise by at least 200 feet. The effect on New York City, for example, would be that even at *low* tide the waters would reach the twentieth floor of the Empire State Building. Virtually all the lowlands of our planet, including its most fertile farmland and some of its densest centers of population, would be inundated.

The other two possibilities would be both swift and appallingly catastrophic. It could be argued, and with some justification, that to cast asteroids into our oceans the aliens would somehow have not only to pull them out of their orbits between Mars and Jupiter but also to haul them from there to Earth—no mean undertaking, at least by contemporary terrestrial standards. To a very highly advanced technological civilization such a feat might not represent an insurmountable problem. They might, on the other hand, find more conveniently placed "asteroidal" ammunition in the shape of the rogue asteroids or "Earth-grazers." These unique planetoids pursue highly eccentric orbits that intersect with those of Earth and the inner planets, Mercury and Venus. The result is that, on occasion, they can and do pass quite close to our world; hence the title of "Earth-grazers." A classic instance of this kind of thing took place toward the end of October 1937, when a hitherto undiscovered asteroid, later named Hermes, was detected by Dr. Reinmuth of Heidelberg. This object approached to within half a million miles of Earth—just a little too close for comfort. By terrestrial standards such a distance might seem immense. Astronomically, however, it is quite short. The mass of this planetoid is considerable; so a collision with Earth would hardly have been a cause for rejoicing. Had it impacted on a continent, total devastation over a considerable area would certainly have resulted. It must always be remembered that damage would be caused not just by the sheer mass of the thing but also by the attendant heat and pressure effects on impact.

One of the more recently discovered (1964) of these jaywalking little worlds has been named Toro. It is about three miles across (these objects are rarely spheres) and can approach to within 9½ million miles of our planet. The consequences of its mass being plunged into the waters of the North Atlantic would be traumatic for the coastal plains and cities both of Europe and of North America. The waves generated would be akin to, or worse than, the dreaded tsunami, which is currently regarded as the most destructive natural phenomena known to man. The most famous tsunami of recent times was that created by the cataclysmic eruption of the volcano Krakatoa in 1883. This volcanic island lies between the islands of Java and Sumatra in Indonesia. The wave raced across the western Pacific ocean at around 500 kilometers per hour (300 mph) raising waves 90 to 120 feet high on the coasts of Java and Sumatra that killed over 27,000 people and swept away everything in their path.

There can be little doubt that if an asteroid fell or was pushed into one of our oceans, a tsunami of monumental proportions would result. It is hardly accurate to term these things "tidal waves" they are totally different in character from the normal tides caused by Sun and Moon. Moreover, the popular idea of a single giant wave traversing the ocean is also incorrect. A tsunami is a *series* of waves, and the effect is identical to that obtained by casting a stone into the still waters of a pond. In the contingency we are considering one of our oceans is the pond and an asteroid the stone. The resulting ripples are, of course, larger. What we get in fact, is a series of enormous waves with the largest crest in the middle of the sequence. Actually, the really giant waves that inundate the shores and land beyond are themselves formed only close to that land, although the cause lies many hundreds of miles out in the ocean.

Compared with normal wind-generated waves, everything about a tsunami or disturbance-generated wave is out of the ordinary. The wavelength (the distance from crest to crest) of sea waves in the Pacific is rarely more than 300 meters. Tsunamis on the other hand have wavelengths of the order of 150 to 250 *kilometers*. Some of up to 1,000 kilometers are not unknown. The same order of difference between wind-generated and disturbance-generated waves is also apparent in velocity.

Wind waves rarely exceed 90 kms/hour (about 60 mph). Tsunami waves can reach 800 kms/hour (500 mph) in the deepest waters of the Pacific.

Another remarkable feature of such a disturbance-generated wave is that the height of the waves in the deep ocean ranges from about three to five meters. The energy locked up in these relatively small waves several hundred kilometers long is tremendous, however, and is retained within them as they race across the ocean. This energy becomes frighteningly apparent as the waves approach land. Their forward motion becomes increasingly restricted, thereby shortening their wave-length and increasing their crest height. Thus a 5-meter wave traveling at 600 kms/hr (375 mph) is transformed into a devastating 30-meter-high wave traveling at 50 kms/hr (about 30 mph) as it heads for the shore.

Because the passage of the deep-water waves is not detectable, the sudden and very rapid growth of a wave near the shore is one of the most terrifying features of a tsunami. A case on record features the crews of several freighters anchored about a mile off Hilo Bay in Hawaii who watched in horror as a colossal wave crest rose slowly up from the sea close onshore to wash away entire portions of the city in a 10-meter-high (30-foot) wall of water. This tsunami had traveled well over 2,000 miles in five hours and was the result of an earthquake in the Aleutian Islands. It was regarded as just a *moderate* tsunami—160 dead and $25 million worth of damage!

Another interesting feature of this type of wave is that the main crest is not the first warning of disaster. Since the main crest is one of a sequence of waves, it is invariably preceded by anything up to seven smaller crests and troughs. These have such an inordinately long wavelength that they look to watchers on the shore like an inexplicable rising and falling of the sea level. In the first instance, a minor wave raises sea level by a meter or two. Within about fifteen minutes this is followed by a trough that apparently within the course of a few more minutes takes the tide out. The heights of the crests and the depth of the troughs steadily increase until the sea retreats well below a normal low tide. This immediately precedes the arrival of the main wave which is generally a huge breaker. Sometimes it takes the form of a bore—an almost vertical step, which represents a 10-to-30-meter-high change in sea level. Bores are believed to form when several of the later waves in a tsunami sequence overtake each other to form a single, highly destructive wave front.

What we have just described are generally the result of seismic disturbances on the sea bed or occasionally of violent volcanic eruptions on island arcs. The even greater effect of a huge meteorite or small asteroid plunging into the ocean can best be left to the imagination.

No doubt the scheme could be enhanced from the alien point of view by plunging these cosmic thunderbolts into our continents as well. The resulting devastation, though no doubt tremendous, would prove less lethal over a large area. The physical effects of the impact might also create rather more material damage than the invaders desired. The aftereffects of inundation would be comparatively less serious since the waters would flow off the land, leaving it ready for Earth's new masters.

The remaining possibility in the field of artificially induced inundation of our planet would depend on whether or not the invaders from the stars had the power, in the form of gravitational beams or the like, to raise the level of the oceans as their starships orbited the Earth. If feasible, that would almost certainly represent a quicker and more streamlined technique than hauling asteroids from their orbits and casting them into the waters of Earth. Each region of our planet could be "processed" at will and for whatever periods of time regarded as necessary.

One possibility that does not seem ever to have appeared in the pages of science fiction (at least to the knowledge of the writer) is the converse of melting the polar caps, i.e., freezing the oceans. Whether or not it is possible for an alien race to wage cryogenic warfare, to produce "freezing" beams is impossible for us to answer, but it may be as likely as some of the other things we have been considering in these pages. To freeze the oceans, thus changing our climate, would bring about the rapid onset of a brief and terrible ice age that would almost certainly put an end to Earth's population sooner or later—sooner more likely, since in the last half century our society has allowed itself to become appallingly vulnerable to the effects of harsh winters which impede the steady flow of fossil fuel supplies, thereby rendering the all-important (one might even say omnipotent) power stations useless. How long could our civilization endure with roads, airports, waterways, and railroads blocked, heat and energy supplies cut off, food running out, and temperatures plummeting? Compared to inundation by water, physical damage would be relatively negligible. The process might take a little longer, but presumably the invaders, safely and comfortably ensconced in their orbiting starships, could afford to wait.

The effects of this type of action would be far removed from those of a mere protracted blizzard and subsequent deep freeze. The deep freeze in this case would be one in which the thermometer plummeted to depths never before experienced by mankind on our planet. Temperatures would be akin to those occurring in the great ice ages of the recent geological past. All that would be missing by way of a direct comparison would be the presence of the mammoth, the wooly rhinocerus and the sabre-toothed tiger! In the circumstances a normal winter would seem almost like a heat wave!

Cryogenics, or low-temperature physics, is an area of research and study that is coming into increasing prominence. The discovery of techniques for reaching and sustaining very low temperatures has opened an entirely new frontier in the study of matter. Present work in the field is centered around the use of external magnetic fields. Heat is created by the random motion of molecules; so the problem is basically one of reducing this motion. In the light of our progress thus far in cryogenics, we would confidently expect the attainments of an advanced technological race in this field to be considerable—perhaps to the extent of producing low temperature effects from a distance.

We have considered the results on the climate of freezing the oceans and said a little about the pronounced and drastic changes that these would involve. For all we know our alien attackers could have several other forms of "weather warfare" up their sleeves, all of it highly unpleasant to the peoples of Earth. The climate of our world, despite its many vagaries and extremes, is comparatively stable. It does, of course, break loose at times, often with dire results. Hurricanes, typhoons, tornados are all fairly typical examples but fortunately these are restricted for the most part to fairly well-defined regions of this planet and even there they are the exception and not the norm. This stability, however, is only a seeming thing. Our climate is in fact a rather delicately balanced entity, and any actions bringing about a change in the established pattern might have entirely unforeseen and wholly disastrous consequences. Thus far, we on Earth have had the good sense to leave the natural order of things alone. It is possible however, that the reason is simply our inability to effect any changes in the weather pattern, rather than sanity of choice. Whatever the reason or motive, apart from seeding clouds with silver iodide crystals to induce rain (with rather indeterminate results) we have left the weather alone. This aspect of future warfare is discussed in more detail in chapter 11.

By disturbing the pattern of the weather seriously through methods beyond and unknown to us, aliens might well bring out cyclonic tempests of unprecedented fury. These are only passing thoughts. Let us hope fervently they will remain so!

8. THE HEAT RAY

Serious consideration of the types of weapons likely to have been developed by technologies centuries or millennia in advance of our own is a somewhat difficult exercise. No matter to what extent one embroiders and embellishes the possibilities, the dominant fact remains that speculation has to be given more or less free rein. Extrapolation into the far future is relatively easy for the writer of science fiction; he is not required to delve too deeply into technicalities and generally ignores detailed explanation. For example, how often is our old friend the "space warp" described in any detail? The thing is generally hidden in a verbal mist that is one part science and nine parts pseudoscience. Perhaps it is as well. The famous French writer of scientific romances, Jules Verne, was one who went out of his way to provide explanations. Though a gifted writer, he was no scientist and consequently his reasoning (though highly ingenious) was, at times, grossly unscientific. Witness, for example, his launching of three astronauts on the first journey around the Moon by means of a monster muzzle-loading cannon, known as the "Columbiad," set into a hillside near Tampa in Florida. The effect of such an experiment on the occupants in real life would have been about as unpleasant (and messy) as it is possible to imagine. The human frame simply does not take kindly to violent and sudden acceleration. Verne's contemporary, H. G. Wells, on the other hand, being a scientist as well as a writer, proposed an even more fantastic scheme—a voyage around the

Moon and back by virtue of a material named "Cavorite" after its inventor, which was capable of negating Earth's gravitational attraction. Such a material is, to say the very least, most improbable, a fact well enough known to Wells. But unlike Verne he did not try to explain the inexplicable. What he did, and that most skillfully, was to weave a web of pseudoscience around the well-established physical fact of refractive index. No one is really convinced by the explanation but without a doubt it sounds highly convincing.

The ordinary nonfiction science writer trying to don the mantle of prophet and soothsayer has to contend with more serious and trying problems for what he envisages must have some real scientific basis either in fact or in legitimate extrapolation of theory. The pitfalls are obvious—which doesn't mean they are always avoidable. Suppose an observer of the Wright brothers' memorable first flight at Kitty Hawk had been given the assignment of foretelling what aviation would be like seventy or so years later. Had he envisaged the wide-bodied jet or the supersonic transport he would have been absolutely correct. He would also have been laughed to scorn by his contemporaries at the time. Had he merely enlarged the Wright brothers' frail biplane into some bigger, stronger thing with umpteen engines and several sets of wings, chances are he would have been considered a true visionary even though his projected creation might be more akin to a flying bird-cage.

Rays of one sort or another have always seemed to epitomize either our future or the contemporary technology of alien beings on other distant worlds. In a century when the scope and concept of radiation has been steadily enlarged, this was bound to seem a valid extrapolation—which indeed it probably is. It is doubtful, though, if ray warfare is going to be quite as colorful and polychromatic as the artists responsible for many of the more lurid pulp sci-fi magazine covers seem to imagine. The number showing spaceships emitting, or being disintegrated by, brilliant, sharply defined beams of red, yellow, blue, green, and purple are legion.

Beam warfare has the capacity for covering a fairly wide field since different parts of the electromagnetic spectrum could conceivably be involved. The genesis of the thing is probably already with us in the shape of the laser, a device that has made, and is still making, enormous strides since its invention a couple of decades back.

Before we attempt, however, to consider these more sophisticated devices, it is probably advisable to think in terms of a simpler facet of the idea—a heat beam using the concentrated energy of the Sun's rays.

Such a beam is certainly not a new concept. On the contrary. Archimedes is credited with the construction of a "burning mirror" during the seige of his native city, Syracuse, in 212 B.C. By means of this device the Roman ships are said to have been set on fire "when they came within a bow-shot of the city walls." Here, perhaps for the very first time in history, is an example of a destructive force being exerted at a distance, albeit a relatively short one.

A heat ray was the principal weapon in the armory of H. G. Wells's octopuslike Martians in his *War of the Worlds,* though the heat source in this instance was apparently not the Sun. Indeed, in some respects, Wells may, without realizing it, have predicted the advent of the laser. Here was a beam that displayed coherence; that is, its rays did not diverge, being instead confined and focused in a narrow, tight shaft. The effect against heavy artillery and warships was absolutely devastating.

The present writer first encountered the idea of focusing the Sun's rays destructively in a boys' magazine of the early thirties. The users of this unusual weapon (the story had a contemporary setting) were the Incas, the original inhabitants of South America occupying that part of the continent now known as Peru. According to some accounts, the Incas did use mirrors to focus the Sun's rays and this may have been the reason for the author's choice. It is certainly well established that the Incas did have considerable skills in building, ceramic, and metallurgical arts, these being attested to by murals and other remains. The Incas were also worshippers of the Sun, so that the use of its rays as a weapon may have seemed a valid and appropriate corollary.

The power of the Sun's rays when focused on a combustible

material is quite intriguing. Indeed, after writing these last few lines the writer decided to try this out in a simple though certainly not original experiment, using an ordinary 2½-inch diameter magnifying lens. When the Sun's rays were focused steadily on the same spot on a sheet of paper, a small hole was burned through it in about two minutes. If focused instead on an upturned corner of the same sheet, the paper eventually smouldered, then burst into flame at that point. In the same way the rays of the Sun, when focused on bare skin, soon become acutely and unbearably hot. That part of the experiment is not recommended to readers!

All this should hardly be surprising, for it is well known that many domestic household conflagrations have resulted from this very cause. At one point during its daily passage across the sky the Sun focuses briefly through a glass ornament or the like onto some combustible material. The result can quite easily be a quick visit from the local fire department. In the selfsame way, many serious brush and forest fires have been caused by the action of the Sun's rays on pieces of broken glass lying on the ground.

We might then justifiably consider the possibility of alien attackers using a system embodying a heat ray. The source of the heat could be either the Sun or alternatively something of an indigenous nature. They could then direct a concentrated beam of pure heat upon selected targets. The weapon would certainly have a most considerable potential.

Ability to utilize the rays of the Sun rather than an indigenous heat source could conceivably simplify construction of such a weapon, since the problem then becomes one largely of optics and insulation. Such a beam could be usable both in and from space, but not, of course, in that portion obscured by Earth's shadow. Neither could it be brought to bear from the surface of the Earth during the hours of darkness or under cloud cover. In these conditions an indigenous source of heat would be essential.

There are three principal modes of heat transfer—convection, conduction, and radiation. In convection, heat is carried or conveyed by the motion of heated masses of matter (as in, "central" heating of buildings), by piped circulation of steam or hot water. In conduction, heat is transferred by contact between contiguous particles of matter, being passed from one to the next without visible relative motion of the parts of the body, for example the passage of heat through the metal plates of a boiler from the fire to the water inside.

Neither of the first two, convection and conduction, has an immediate bearing on a hypothetical heat ray weapon, the like of which we are evisaging. The third, however, quite definitely has. In radiation, the heated body, through the thermal vibrations of its atoms, radiates vibratory waves into spaces that are electrodynamic in nature. If the emitting body is at a sufficiently high temperature (e.g., red-hot), these waves are partly of visible light. Should the body be at a lower temperature, radiation is confined to the infrared portion of the spectrum, which is outside the visible region. Infrared radiation is similar in nature to light but contains sufficiently less energy per unit (photon) to render it incapable of exciting the optic nerve and so being perceived by the human eye. Such waves can be absorbed by matter on which they impinge where they are reconverted into heat.

The three modes of heat transference are subject to different physical laws. In space, the medium in which we are most interested, it is *radiation* that is paramount. Indeed, it might be wondered how, in a vacuum, convection and conduction could have any relevance whatsoever. The reason is that, despite seeming evidence to the contrary, space is not an absolutely perfect vacuum. There is an old saying that nature abhors a vacuum. Even in space there is an element of truth in this.

If the radiation stream inside an impervious enclosure at a uniform temperature is independent of the nature of the walls of that enclosure and is the same for all substances at that temperature, it follows that the full stream of the radiation in such an enclosure is a function only of the temperature—or, in more colloquial terms, the higher the temperature the stronger the radiation so long as certain other essential physical requirements are satisfied. For example, radiation from a white-hot platinum wire at $1,200°C$ is 11.7 times that of the same platinum wire at

525°C. With this in mind we can envisage certain possibilities. Suppose, for example, that the heat from a small, confined thermonuclear reaction could somehow be focused and used as a weapon. So far, and perhaps not surprisingly, our technology has failed to come up with an answer to containing and controlling such a reaction. At present no known material can remain solid at the exceedingly high temperatures involved, and such progress as has been made to date has centered on magnetic confinement of the plasma. The extent of the problem will be fully understood when we say that the temperatures involved range from 45,000,000°K to 400,000,000°K.

To control nuclear fusion (i.e., thermonuclear reaction) the reacting system must be confined. It is doubtful if even the most technologically advanced race in the galaxy (or in any other) can have devised a *material* with the required degree of resistance. What they could have done is to proceed much farther down the road *we* have already begun to tread—confining the plasma magnetically. This seems a reasonably feasible possibility, though it is obviously one fraught with many great difficulties, and no doubt perils as well.

A heat ray involving temperatures of this order would have some measure of claim to being the ultimate weapon—the ultimate both in effect and horror. The mere touch of such a beam would instantly vaporize anything. Against it there could be little defense. Rocks would be reduced to pools of bubbling magma, metals to streaming white liquid, the oceans to steam, human bodies to vapor.

A few paragraphs back we touched on the utilization of solar energy as the heat source for such a device. This would certainly eliminate the necessity to confine the almost unconfinable. On the other hand, such a system would be much less mobile, and would generate far less heat, thus greatly reducing its lethal powers as a weapon. In space the Sun is always around and here, of course, is a thermonuclear pile of immense and virtually unending capacity. The problem in this instance is largely one of optics, and more easily surmountable.

No details appear to exist of the heat ray alleged to have been used by Archimedes against the Romans during the long siege of

Syracuse or just what degree of damage was done to the Roman ships and their occupants. Presumably a highly polished surface of suitable shape and configuration reflected and focused the rays of a hot eastern Mediterranean sun onto the sails and dry, tarry timbers of the vessels. Perhaps only one or two actually caught fire. Panic could then have done the rest, since a fire ship among others is a somewhat unenviable prospect to all concerned. Whatever the circumstances, it must have seemed to the Romans that the defenders of Syracuse had developed some uncanny power. A much more recent experiment carried out by G. L. L. Buffon in a Paris garden in 1747 showed that the feat attributed to Archimedes back in 212 B.C. *could* have been accomplished. On this occasion sunlight reflected from 140 flat mirrors ignited a stack of wood placed about 200 feet distant. The renowned French chemist Lavoisier refined the process considerably by enclosing specimens of various substances in transparent quartz vessels and placing them at the focal point of a 52-inch diameter lens. The resultant incineration due to highly concentrated solar radiation appears to have been most effective.

In fact terrestrial man has endeavored for many centuries to make direct use of the Sun's abundant radiant energy. On the whole, any success has been rather meager, largely because of the fact that sunshine is intermittent, variable in direction, and relatively low in intensity. It was found that apparatus designed to utilize solar energy had to be fairly extensive in area and have an inbuilt facility to "store" energy for use during periods when the Sun's rays are not available.

The earlier endeavors of Archimedes, Buffon, and Lavoisier are reflected today in the modern solar furnace. In this appliance parabolic reflectors, similar to those used in searchlights, are able to concentrate solar radiation so effectively that temperatures as high as 3,500°C have been attained. Equatorial mountings akin to those used on large astronomical telescopes are employed to enable the mirrors to follow the apparent daily motion of the Sun across the sky. It is clear that the rays of radiant heat energy from the Sun *can* be utilized and to a considerable degree. This may yet be a fact of vital importance to mankind once our limited stocks of fossil fuels are exhausted.

Only about 46 percent of incoming solar energy actually reaches the Earth's surface. Of the rest, 35 percent is reflected back into space by clouds and the remaining 19 percent is absorbed by the atmosphere. It follows from this that similar experiments carried out in space above the atmosphere would prove over twice as effective. Sophisticated solar beam devices embodied in weapons might therefore be used against defending terrestrial spacecraft and, if conditions permitted, perhaps against targets on the Earth's surface. It is unlikely however, that a solar-type beam could ever be as lethal as one containing its own "pocket Sun" i.e., a controlled thermonuclear heat source. As far as we are concerned, such a weapon belongs only to science fiction. Nevertheless, there exists the disturbing feeling that such a weapon, despite enormous technical problems, could eventually become a feasible proposition here on Earth. If time is the only other factor upon which the emergence of such a terrible weapon is dependent, then eventually it must appear on the scene. On that day a new and even more horrible dimension will have been given to warfare. Adequate time is already what other civilizations within the galaxy will have had to perfect such a device. With these thoughts intruding, one begins to realize that what could be dismissed as fiction a few decades back could eventually come to pass. In 1906 H. G. Wells envisaged the atomic bomb. Less than forty years later fiction had become horrible fact. Let us hope that what we envisage today will not also be with us in forty years—or less!

Already brief (milliseconds) bursts of contained thermonuclear energy have been attained (the Zeta experiment carried out in England in 1953); already nondivergent (coherent) beams have been perfected in lasers. Somehow link the two and add the essential ingredient—time. In the case of other galactic civilizations add interstellar capability and a love of conquest. The result could be a power beyond anything yet imagined—or imaginable.

For every poison, it is said, somewhere is the antidote; and for every means of attack, a means of defense. In this particular instance we have probably come upon an exception. A beam as hot as the interior of the Sun itself is not something likely to be stopped or deflected by anything, save perhaps an opposing heat beam that strikes the opposing projector first. Some may say such a weapon is impossible. There were those who said that about the atomic bomb. Even among those waiting to witness the first nuclear detonation at Alamogordo, New Mexico, in July 1945, were a number of scientists still highly sceptical about the result. There are some highly dangerous words in our language. "Impossible" is one of them.

9. STELLATOMIC WAR

When in the preceding chapter we spoke of a thermonuclear device acting as a heat source or generator for a devastating heat ray it may have raised in readers' minds the question of straight-forward atomic attack in any future stellar war. Clearly something along these lines must be a very strong possibility, since nuclear energy is after all the fundamental force of the entire universe. It is by reason of nuclear energy that the stars shine and, since galaxies comprise millions upon millions of stars, it is also by this means that galaxies shine. When a thermonuclear device explodes, we are witnessing in a very small way, and for a brief interval of time, the basic process that keeps the sun and all the other stars in the heavens shining—the transition of hydrogen to helium with the accompanying production of vast amounts of energy. It is an efficient method of power production and could we harness it for use in our power stations, as someday we will have to, if our civilization is not to go into decline, it would make power production by the burning of fossil fuels seem incredibly inefficient and crude. Nuclear energy must be the force of the future; it must be the force that drives the technologies of advanced cosmic communities throughout the galaxy. It is consequently a force that has to be considered in the broad field of advanced weaponry.

Nuclear weapons are, of course, something with which we on Earth are already only too horribly familiar. Nuclear energy harnessed for peaceful purposes could be the savior of our civilization; nuclear energy geared to war, its end. Even now the entire human race is at risk from the weapons already developed and deployed by fellow beings on our own small planet. In any future global conflict those not actually killed by the incredible heat and blast of nuclear weapons would, as likely as not, fall victim to lingering death because of fallout and radiation, leaving Earth a lifeless, charred cinder.

Since we have, in the past few decades, produced a selection of extremely powerful nuclear weapons, there might seem little more that any other cosmic race could accomplish in this field. Already we have attained a high measure of overkill potential, to use the horrifyingly cold-blooded jargon of nuclear war. Certainly, in the megaton-range, thermonuclear devices developed in the past twenty years by the superpowers, we have weapons that can totally and most effectively devastate very large areas through heat and blast effects alone. Races our technological superiors could easily have extrapolated on this considerably, producing bombs and projectiles capable of devastating entire continents. Indeed, it may be safe to assume that at various points throughout our galaxy there exist what we could describe as the awful warning to other civilizations—lifeless, radioactive planets whose peoples played with Promethean fire until they got burned. Such burns do not heal. They are fatal—and fatal is forever!

There is another way of looking at the position with respect to an attack on us from interstellar space. We have said already that another race wishing to take over our planet for their own purposes might prefer not to devastate its surface too completely. Neither would they want to leave it dangerously radioactive. Nuclear weapons capable of laying waste an entire continent might also damage the relatively thin crust of the planet, causing tremendous outpourings of magma from the mantle. The crust of the Earth is only about 35 kilometers thick so far as the continents are concerned. Beneath the sea this shrinks to 5 kilometers. The risk is therefore real enough, especially near extensive and deep geological faults. Our alien foes might consequently have serious misgivings about casting their most powerful and devastating nuclear projectiles against us.

Destroying the indigenous population of Earth could be one of their aims; laying waste its surface or damaging it structurally would probably not fit in at all with their longer-term plans.

At this point in the argument there is one type of nuclear weapon that immediately suggests itself—one that has the capacity to kill, but destroys only to a very limited extent. It is one already envisaged here on Earth, though as yet still not developed—the so-called neutron bomb. Here indeed is a weapon highly appropriate to a race wanting our world but not us. In a year or two it is virtually certain that the neutron bomb will have found its way into the arsenals and stockpiles of the world's great powers. Whether this will advance or retard the chances of Armageddon here on Earth only time will tell. Presumably it will join the ranks of the other deterrents, and our fragile civilization here on Earth will continue to creak along in an uneasy armed peace. Once two potential adversaries have the weapon there is every possibility that it will not be used, as has happened with the atomic and hydrogen bombs. If one side hurls it, the other promptly returns the compliment with interest so a balance of power—or terror —results. But aliens sweeping omnipotently around our world, perhaps with their starships protected from terrestrial missiles by force fields, could easily, methodically, and with considerable malice aforethought sow a pattern of neutron bombs liberally over every continent and island. And there would be precious little we could do about it. Mankind, all of mankind, would very soon be dead, but his cities, airfields, ports, roads, mines, factories, and other essential artifacts would remain relatively undamaged and ready for alien use if required—all very neat, efficient, and tidy.

Neutron attack could then be a form of warfare greatly favored by another civilization in the galaxy wanting our world for its own. But since we are already on the point of developing the neutron bomb, is it not possible that a race competent to reach us from the stars, perhaps by transit through an unknown unsuspected dimension, would have refined neutron weapons very considerably? Our original neutron bombs and their developed successors might seem to them incredibly crude, just as today, only four decades later, the original fission bombs dropped on Hiroshima and Nagasaki are regarded as museum pieces.

Since the prime purpose of the neutron bomb is to kill by radiation, it is presumably this property that would be subject to development, perfection, and sophistication by cosmic adversaries. The bomb or projectile explodes and untold millions of deadly subatomic particles are released. The lethal potential is sophisticated enough, but the method of dissemination is comparatively crude.

In the preceding chapter we considered a ray or beam of concentrated heat. Would it be extrapolating things too far to attribute to aliens having a highly advanced technology the ability to direct a beam of neutrons or other damaging subatomic particles toward their adversaries? Here perhaps is the death ray so beloved of the earlier science fiction writers. Certainly it is a possibility that does not seem too far-fetched, one not incapable of realization.

Not only could a *neutron ray* be used against population centers on the surface of the Earth, it could serve equally well against terrestrial spacecraft moving in to attack the invaders. Caught in beams of such lethal propensity the ships of Earth would very soon contain only sick and dying crews. At present, death by this means is supposed to ensue in a day or two, but such a degree of concentration could have been developed that the effects of the beam, if not instantaneous, might prove lethal within minutes.

Here, of course, is the essence of a possible defense that, if developed by ourselves, might stand us in very good stead against any attempted alien takeover of our planet—assuming, of course, that the metabolism and physiology of the aliens were susceptible to an attack of this kind. It is just possible that what might so effectively and horribly kill us might have no effect on them or, even more ironically, prove positively beneficial. Since biology is unlikely to vary too much, however, it is probably a reasonable assumption that a concentrated beam of this nature would not exactly improve their health!

A more likely prospect is that they would in some manner

have protected their vessels against such subtle forms of attack either by the use of specially shielding materials or external deflecting force fields of one sort or another. The latter have long been a favorite in the starships of science fiction. The pity is that the authors concerned do not tell us how these force fields are achieved.

The use of very unpleasant varieties of fission bombs and projectiles in an interplanetary war of the future is undoubtedly another alarming possibility that could never be wholly discounted. Already we realize how lethal would be the results of any war employing that bloodcurdler of not so very long ago— the so-called cobalt bomb. This charming device would, were it created, spread around a particularly nasty radioactive isotope of cobalt, which would endure for ages. Weapons producing harmful and extremely poisonous radioactive materials with very long half-lives could, however, prove a most serious hindrance to the victor in any atomic-type war. If Earth is the prize, then presumably it is only an Earth in relatively unspoiled condition that would be acceptable. A planet the surface and atmosphere of which had become contaminated and poisoned by materials that could continue to display their deadly effects for perhaps thousands of years is hardly likely to prove acceptable to the victors. Cosmic attackers would be well aware of these facts and this we will hope downgrades the possibility of this particular form of horror. Let us, in closing, also hope and trust that the men, now and in the future, who guide the destinies of *this* little planet are equally aware of these facts. As things stand today we should perhaps go more in fear of some of them than of aliens.

10. ATTACK BY STEALTH

So far we have been contemplating attack in the best science fiction traditions—a full scale alien descent upon our planet using some very awesome weaponry. Perhaps now for a while we should think along slightly different lines.

The possibilities here can probably best be illustrated by studying a hypothetical situation. Suppose that sometime within the next century a large, and very obviously alien starship were to approach our world and then proceed into a parking orbit around it. The occupants of this mysterious vessel make no apparent move. It emits no radio signals; neither does it respond to any from Earth or from terrestrial space vehicles that approach it. At no point does it make any overtly hostile move. In fact it makes no move at all. Not unnaturally the terrestrial spacecraft are reticent about commiting anything that might be remotely construed as a hostile act. With this great futuristic creation sweeping splendidly around our world, this is clearly understandable and undeniably prudent.

By now all the governments of Earth are painfully alive to the inherent dangers of the situation. Space forces of the combined nations are on full red alert, stationed at a discreet distance with all weapon systems at the ready. As yet, military preparations on the surface of Earth are less organized. The threat is up there in space. If attack and invasion portend, they will come from that quarter. And it is there that they must be stopped. Alien forces must not be permitted to reach the surface of our world.

Every day as the sun sets, its rays illuminate the super starship in its ominous, silent orbit high above Earth. All over the world people stop whatever they are doing to gaze at this peculiar and unreal spectacle. Many are strangely troubled, some increasingly so. Here at last is something from the stars, from another and obviously much greater, infinitely more advanced civilization. And in the mornings when the Sun rises it is still there—huge, magnificent, menacing.

More and more terrestrial craft join the great vigil in space. Now it is not only war craft, but shuttle ships full of sightseers or newsmen. The strange visitor has become big business as well as a threat. The passing of time only serves to heighten the drama being played out hundreds of miles above the Earth's surface. But still the enigmatic vessel from the stars remains aloof and silent. Does it really contain occupants, intelligent reasoning beings of unknown form and, if so, what are their intentions? The security screen of Earth forces moves closer and tightens. All orbiting space stations are on continual alert. The leaders of the world confer, argue, and hope, but even as they do, a great army of alien beings with a panoply of weapons, irresistible and beyond human understanding, materializes silently and efficiently and with ruthless intent on the relatively unguarded surface of Earth! They have descended in force from their great starship, yet not one of the host of defenders has seen them pass. No ports or entrances on the vessel have opened. No vestige of activity has been seen. An invasion task force from the stars has descended upon Earth apparently out of thin air. With grim and ruthless efficiency it sets about its terrible purpose.

"Beaming down" to the surface of a planet from a space vehicle high in orbit above it is hardly a new idea to devotees of science fiction. It is one that has been popularized both in the United States and in Europe in recent years by the enjoyable television series "Star Trek." Captain Kirk, Mr. Spock, Lieutenant Uhura and others of the redoubtable crew of the starship U.S.S. *Enterprise* have been doing this efficiently and apparently quite painlessly for some time now. In fact, viewers of the series by now regard the operation as perfectly normal. It is certainly an easy and quick method of transit and saves messing about with shuttle craft and other sundry forms of crude paraphernalia. It also saves screen time and lets everyone concerned get on with the plot.

As we have already seen in our hypothetical glimpse into the future, a technique of this sort in an invasion of our planet (or of any other inhabited planet) would bestow very considerable advantages on the attackers. Without this, the invasion force would either have to take their starship down to the planet's surface, a highly hazardous undertaking, or release scores of shuttle craft. This would inevitably prove a slower business, the essential element of surprise would almost certainly be lost, and disembarkation could prove vulnerable. If, on the other hand, they can somehow pass invisibly from ship to planet and there rematerialize, the defenders, who would probably not stand very much chance at best, could be taken completely by surprise and be totally demoralized from the very start.

At this point we must stop and consider, if we can, the practicalities of matter transfer. The big sticking point is that of dealing with something that, so far as we are concerned, does not exist. In the first instance, a simple and familiar analogy may be useful. If we are desirous of sending a photograph or important document to another part of the world, it is no longer necessary that we send the actual item. It may nowadays be reduced to a series of black, white, and gray spots that are then converted into a fluctuating electric current that passes at the speed of light (186,000 miles per second) to its destination. So far as points on the surface of the earth are concerned, transit time is, to all intents and purposes, zero. At the receiving end of the link, the picture or document is reconstituted into an exact facsimile of the original. Might it therefore be possible to reduce matter in an analogous way and transmit it in like manner? This would, somehow or other, entail its conversion into a fluctuating beam of photons, the minute "particles" (for want of a more appropriate term) that constitute light and other forms of radiant energy, which would then be transmitted across space for reconstitution at the receiving end. To achieve something along these lines with small, inanimate objects would be a very considerable and marvelous feat; to achieve it with a human body or other living organism would virtually border on the miraculous. The transmission of matter, animate and inanimate, would be an accomplishment of great worth and almost unimaginable complexity. At the present point in human affairs, our technology is totally unable to come to grips with the concept. Can we then attribute such powers to alien beings in parts of this galaxy and of others? If they are sufficiently far ahead of us technologically it is, we must suppose, possible that they have developed a technique of matter transmission. Obviously we cannot be categorical either way. We must not assume that, given sufficient time, technology can achieve virtually anything. There must almost certainly be feats forever impossible to accomplish, irrespective of technological prowess and expertise. Nevertheless, matter transmission does seem like something that probably could be achieved even by ourselves some day—though that day undoubtedly lies very far in the future.

The possibilities for inanimate matter transmission can probably be regarded as brighter than those for the animate variety for a number of reasons. The human body might conceivably respond in the same way as a nonliving object. The treatment, however, would presumably have to be regarded as something less than successful if what was reconstituted at the receiving end was merely a fresh, complete, and undamaged corpse! Even if the subject of the exercise did not complain, his or her relatives might feel rather aggrieved! Presumably experiments in this highly risky field would first be carried out on animals. Even if these experiments were 100 percent successful, it would indeed be a brave man or woman who was willing to try out the new technique personally for the first time. As an experience it would certainly be of the most unique kind and what the sensations and feelings would be are quite impossible to suggest. It could be like a moment's sleep or brief instant of darkening.

Accidents resulting in the delivery of only a corpse might be preferable to any in which the subject of the experiment arrived still alive but improperly reconstituted, grossly deformed, or with mind totally deranged. A few years ago a science fiction movie fastened on this alarming theme. It so happened that in the transmission chamber, unknown to all concerned, there was

also a fly. The result of this intrusion was a living human body
with the head of a fly! What happened to the other members,
that is to say the head of the man and the body of the fly, was
not revealed. Presumably this would have meant two horrific
central characters, which would have entirely upset the carefully
thought out thread of the story. Neither was it explained how
the head of the fly had grown to the proportions of the human
head or how, with such an impediment, the man was still able to
speak and reason coherently.

We might not be called upon to suffer misfortunes or indig-
nities of this sort, but there does seem a horrible potential for
things to go awry in a host of bizarre and alarming ways. If a
number of objects were processed at the one time, there would
have to be a firm guarantee that on reconstitution at destination
point each comprised only its own material *(and all of it)*. The
guarantee would be especially vital if the objects concerned were
two or more human beings! We can only assume that if matter
transmission *is* possible and if certain alien civilizations in our
galaxy have the ability to carry it out, such difficulties will have
been totally eliminated.

In our hypothetical episode earlier we were thinking only of
aliens beaming down to the surface of our planet from a starship
in orbit above it. But surely, it can be asked, if this becomes
possible, couldn't it be extended in such a way that starships
themselves become redundant? Why not simply transmit the
invasion force from its home planet to ours? It would be
cheaper, quicker, and simpler—or would it? The photon beam
that is an alien in the process of being transmitted to Earth from
a planet orbiting our nearest stellar neighbor, the star Alpha
Centauri, takes 4.3 years to reach us. Traveling at the speed of
light, the journey will have seemed timeless to him despite the
passing of these 4.3 years—or so we will suppose. If the two stars
(Sun and Alpha Centauri) had been instead 1,000 light-years
apart, then a millennium would have rolled by in what was still
to him an instant. Thus, where long interstellar distances are
concerned, the time element is still inconveniently and unacceptably long. What happens if the alien now wants to return?
Unless he has had duplicate transmitting apparatus beamed

with him he cannot. It is as simple as that. And if he has
brought along the means of return, he goes back to a home
planet 2,000 years older where he would by then not even be
remembered. At a separation of 4.3 years, however, the latter
problem is less acute.

We know also that radiation is subject to the inverse square
law, which is merely another way of saying that it becomes
greatly attenuated by distance. It could therefore be decidedly
hazardous for living beings to allow themselves to be transmitted over very great distances. They might arrive at the receiving
point in an extremely anemic or haggard condition. In fact, they
might not arrive at all, and remain forever as a long drawn out,
weary string of photons shambling pathetically through space.
This would certainly constitute an enduring monument of the
most novel kind!

The main inherent weakness in devices for matter transmission as they generally appear in science fiction books and movies
is the absence of receiving equipment to reconstitute the transmitted item, article, or body. In "Star Trek", as in so many
others, the crew members being so conveyed materialize on the
surface of the planet quite literally out of thin air. There is no
device there for their reconstitution. And when they return to
the U.S.S. *Enterprise* from the planetary surface, no transmitting
apparatus is present to perform the highly essential function. All
is inexplicably controlled by apparatus within the great starship
itself. This is all very neat and convenient. Such a capability is a
tremendous extrapolation of the whole concept of matter transmission. Though it remains well beyond our contemporary
powers, we can at least appreciate how in some manner matter
could be converted into radiation, transmitted, received, and
reconverted to matter. It is much less easy to appreciate a technique that dispenses entirely with the need for receiving apparatus. It is, in fact, roughly analogous to sending a radio signal
from a transmitter and expecting people at a distance to hear it
without the necessary radio receiver.

The hypothetical situation envisaged at the beginning of this
chapter was admittedly based on this highly extrapolated premise. Obviously, matter transmission without the necessity of re-

ceiving apparatus "in situ" at the destination point possesses a very clear advantage, especially in the military field. If potential alien invaders of our planet wish to reach its surface by stealth from an orbiting space vehicle and positioned receiving apparatus is mandatory, then a few of them must first somehow slip down unobserved and surreptitiously construct the necessary receiving chamber. Should it be destroyed or seriously damaged by Earth's defenders, the invasion might have to be abruptly terminated, leaving several members of the task force and their equipment as bundles of photons somewhere between Earth and ship—a uniquely horrible and probably well-deserved fate!

The seeming necessity for some form of receiving and reconstituting apparatus at the destination end would appear to downgrade the whole concept of matter transmission in a military context, although after a bridgehead had been established presumably further reinforcements and supplies could be beamed down and reconstituted by the apparatus set up. If, on the other hand, the necessity for receiving apparatus could in some way be eliminated, the scope of the technique becomes tremendous. Should it be possible to give it interstellar range as well, even if only of the order of a few light-years, then possessors of such apparatus could well, in time, control much of the galaxy. When we examined the possibilities of artificial black hole transit through space, the threat of a "space tunnel" opening up near our world was stressed. Here now is a different kind of space tunnel which could open on the very surface of our world. We would not see it—only the creatures that emerged from it. What we saw might be far from pleasing to our eyes, and the consequences highly detrimental to our future.

11. ENVIRONMENTAL WARFARE

In a sense this chapter should be seen as an extension of chapter 7, in which we considered what might be achieved by alien attackers if they chose to raise the level of our oceans by melting the polar ice caps. We also looked briefly at what could follow the freezing of the oceans and the devastating effect of giant waves produced as a consequence of casting small asteroids or very large meteorites into the oceans. In short, we were concerned for the most part with actions that, in one way or another, involved the oceans. There are, however, several other facets of environmental warfare that might also be possible, and in the most unpleasant of ways. Already here on Earth actual techniques have been suggested (and in some cases investigated) for modifying the weather, the ionosphere, even for triggering earthquakes, as a means of inflicting large scale death and destruction. The philosophy seems to be why use hardware to destroy a foe if somehow nature can be persuaded to do the job instead.

Weather modification as a weapon comes in several forms and does not merely consist in altering the rainfall pattern. It includes modification of fog, hail, storms, and even lightning. So far as we are concerned at the present time, such things are still theoretical. To a vastly superior intelligence they could very easily constitute a practical, highly effective method of waging war.

Fog, despite our attempts to cleanse the atmosphere of smoke and fumes, is still an all too common phenomenon and one that can very easily tie up transport services, especially aviation. It is not then hard to imagine how convenient fog might be to an enemy in time of war—infinitely more so could that fog be producible at will by the enemy. This could be particularly valid were that enemy an already landed alien force. How much easier its task were we "blind" or confused. The octopuslike Martians created by H. G. Wells obviously realized this as they cast about their canisters of dense, black gas. The power of nature and the imagination of a writer are, however, two entirely different things. The latter, in this instance, was invoking the use of what was really only a super smoke screen with an additional lethal potential. Smoke screens of one sort or another are often used on modern battlefields and in naval engagements. The main disadvantage of the technique is that smoke tends to disperse and dissipate itself too easily to confuse or tie up an enemy for very long. Natural fog can be much more effective in this respect so long, of course, as it does not tie up the attacking forces as well. Unfortunately real fogs cannot be brought down at will—or can they?

Fog is formed when moist air is cooled. As the humidity approaches saturation point (the point at which the atmosphere will hold no more water vapor), small droplets of water begin to obscure visibility. This at once suggests a possible means of fog creation; by the removal from the atmosphere of sufficient heat, the humidity of the air is raised to near saturation point. This method has one major drawback; it necessitates very large amounts of energy. But whereas this represents a source of difficulty to us, it might not to an alien invasion force with massive amounts of energy at its disposal. Nevertheless, as a technique it does appear undeniably cumbersome, and so chemical initiation of fog might well be advantageous. Already experiments have shown that in instances where relative humidity is well below 100 percent, fog can be produced by the use of hygroscopic seeding materials. Such materials, because of their pronounced affinity for water vapor, can quickly initiate condensation. So far, experiments along these lines have relied for their success on prevailing wind conditions—the less the better. A strong wind, even a respectable breeze, would soon disperse any fog thus produced, and of course on a day with wind present the fog would simply disperse at once. It would seem then that alien invaders would also have had to develop a certain degree of weather control and manipulation. This might be much less impossible than it seems.

One of the most spectacular of meteorological phenomena is undoubtedly lightning. An electrical storm with its vivid and violent lightning flashes is a most impressive and awe-inspiring manifestation of nature's might. Here is tremendous, naked, raw power going quite literally to waste. If these vicious bolts of lightning could somehow be harnessed and directed against chosen targets at will, a weapon of considerable potential, both physical and psychological, would have been forged. So far as we are concerned, there for the present the matter must end. Probably this is just as well. During the past few decades we have already produced more than enough dangerous and lethal "toys." The difficulties in harnessing the power of the lightning bolt are twofold. The mechanism of lightning is still imperfectly understood. Research and development might eventually take care of that aspect, but the other is less easy to counter. The frequency, extent, and location of electrical storms is something of a chance affair. Violent thunderstorms are taking place at various spots on our planet's surface virtually all the time. It is a safe bet, however, that they would not conveniently come along at the specific place and time required by military men equipped with the means to harness and use their power as weapons against an enemy.

One method of modifying lightning strikes has, according to reliable accounts, been tried with some measure of success. It is based on a sudden perturbation of the electric field, in which lightning is triggered artificially by launching a number of small rockets carrying a thin steel wire, one end of which is connected to the ground. Whatever the degree of success here, the technique seems impractical and cumbersome and the potential for development extremely low. And the fact still remains that we have to wait for convenient conditions. Such a wait could prove quite protracted, especially in some parts of the world where a distant rumble of thunder is about as close as an electrical storm is ever likely to come. Even if alien invaders had developed a highly efficient technique for harnessing and directing lightning bolts, they, too, would presumably have to await conveniently charged atmospheric conditions—unless somehow they were able to reproduce the conditions leading to the evolution of thunderstorm conditions.

Thunderstorms normally comprise several nuclei or "cells," each of which represents a separate electrical entity. These so-called cells have lifetimes of approximately 45 minutes. It is unusual for them to develop simultaneously, one "cell" generally becoming active as another starts to wane.

The mechanism or mechanisms responsible for the creation of atmospheric electricity are complicated, and it is doubtful if they are even now wholly understood. Indeed, there may be several diverse factors at work. The impact of small ice particles and supercooled water droplets upon beads of soft hail are known to produce considerable electrification particularly when usual temperature differences exist between the impinging bodies. Other factors leading to thunderstorm conditions are coalition of charged droplets, mass migration of charge by air flow, and preferential attachment of positive or negatively charged air molecules to precipitation elements. Reproduction of these processes on a locally intense scale might not constitute an impossibly difficult task for a highly advanced technological society. Alternatively, alien invaders might, in some other fashion, generate their own high potential and direct this downward in the form of destructive lightning bolts.

The idea of alien starships sailing serenely through our skies while searing cities, armies, and defending aircraft by means of lightning bolts seems very much like the kind of thing portrayed on the covers of certain highly sensationalized pulp sci-fi. It also has strong connotations of Old Testament wrath. In those distant times lightning struck great fear in the hearts of men and women. Even in medieval times and later, death and destruction by lightning was seen as an act of God. And even now there are those who become quite agitated at the first distant mutterings of an approaching thunderstorm.

Despite its spectacular nature and the damage which at times it can undoubtedly do, is lightning all that dangerous? It is a very brief discharge that may persist for up to a second at most. At times the electrical current involved approaches a value of 100,000 amps, which is certainly not inconsiderable. At others, the figure is around 100 amps. Persons struck by lightning are generally either killed or they recover completely, albeit severely shaken. The lethal current is small and the bodies of many victims show no markings whatsoever; visible burns are present only in the case of close strokes. Lightning nearly always tends to go for the highest points, which is why trees, especially isolated ones, are such frequent casualties. In dry, heavily wooded areas lightning strikes can be the cause of very damaging forest fires. Immediate damage to buildings is generally minor, though much depends on what the building happens to contain. If, for example, its contents comprise high explosives, the results can be both devastating and highly spectacular. This was amply demonstrated on July 10, 1926, when lightning struck a magazine belonging to the U.S. Navy at Lake Denmark, New Jersey. Before the smoke and fumes had cleared away sixteen persons were dead and $70 million worth of property had been destroyed. Debris fell as far distant as 22 miles and all buildings within 2,700 feet of the blast were totally flattened. This, it is reckoned, must have been the most expensive lightning strike in history. If lightning bolts *could* be directed against specific and suitable targets (e.g., explosives magazines, ammunition dumps, fuel storage tanks) their propensity as a weapon would be considerable. Nevertheless, nuclear weapons are infinitely more damaging, and indeed conventional chemical explosives in the shape of artillery shells or aerial bombs could easily achieve anything a lightning strike could produce. The use of a directed lightning bolt might seem, however, to offer a good chance of destroying a selected target while sparing nearby structures. On the other hand, laser beams would be even more precise and probably, by their nature, prove a lot more effective.

In the field of offensive weaponry, much might depend on the way the research and technology of a specific alien community had been oriented. Despite the objections we see they might have elected to produce fantastic electrical machines whereby bolts of high voltage electrical energy could be directed at will against hostile craft, personnel, and installations. And who knows but that they might even have become able to use the terrible power of the storm. *"Eripuit coelo fulmen"* wrote the Roman poet Manilus in the first century A.D.—"He snatched the lightning from the sky." For such a race this would indeed be an appropriate motto.

The deliberate modification of a hurricane, could this be achieved, would undoubtedly lead to a force of tremendous potential. There are two significant factors involved here. First, there is the transfer of latent heat from the surface of the ocean to the air inside the storm. This is absolutely essential if a hurricane is to reach and retain full intensity. Second, the energy of the hurricane is released by convection in highly organized convection "cell" circulations. It is even now considered theoretically possible to use hurricanes and similar type storms as weapons by enhancing, dissipating, or guiding them by means of cloud seeding or other techniques. A hurricane, developed and controlled in this manner, could be used devastatingly against heavily armed and fortified coastal areas.

It is admittedly easy to envisage a synthetic, directed hurricane, even easier just to state that, somehow or other, an advanced alien race could use the elements in this fashion as a highly potent weapon. More validity could be given to the idea were we able to postulate something of the meteorological mechanics involved. A suggestion having this end in view has, perhaps surprisingly, already been made. To generate a hurricane heat transfer from the sea to the atmosphere would be essential. The evaporation from large areas of ocean, it is now believed, could be modified by the spreading of thin layers of oil over the surface of the water, hardly an insuperable undertaking. This technique, in conjunction with cloud seeding, it is suggested, could, in theory at least, be used to direct a hurricane to destroy enemy coastal defenses. This might or might not work. But if hurricane generation and direction is a weapon in any alien armory, surely a more sophisticated, less messy technique would be employed. The use of such a "weapon," even by aliens, must remain highly speculative. At the same time it must be said that a hurricane is a very devastating thing, as residents along the shores of Florida, the Carolinas and the Gulf of Mexico will readily testify. Deliberate generation and control of whirlwinds (or "twisters"), if feasible, could also prove highly devastating. Since these wreak their destructive fury over a fairly narrow path, they could be ideally used against airfields, seaports, and similar installations.

A rather worrisome feature of what we might call "weather warfare" is the fact that such techniques are already being discussed and pondered over here on Earth, for use presumably in future indigenous conflicts on this planet. This is clearly indicated in the words of a recent feasibility report on the subject. "Modification of climate is, on the whole, still in the realms of theoretical possibility" says the report, "but it is envisaged as a strategic weapon which could, for example, be used to destroy an enemy's agricultural pattern. It is thought that it may not be possible to achieve such changes in large scale atmospheric circulation in the coming of two or three decades." Such prospects are not exactly pleasant. Perhaps there are those of us who might already like to quit Earth for some more peaceful star planet if such delights represent the shape of things to come.

Environmental warfare conducted either by aliens against us, or by ourselves against each other, is bound to include the possibility of initiating earthquakes at a time and place most inconvenient to an enemy. Seismic tremors, if sufficiently strong, can be tremendously devastating not only to lives and buildings, but also to communications. Roads and railway tracks can be shattered or twisted out of recognition, airfields totally wrecked, telephone and power lines brought down and severed. All this is bad enough in normal circumstances, but potentially fatal if the area involved happens to be the hinterland of a force defending it from attack. Fortunately (or at least so we presently believe) seismic warfare could only be conducted in certain fairly well-defined areas, i.e., in regions of crustal strain.

There would appear to be two principal methods for triggering earthquakes artificially. The first is by using explosives sufficiently powerful (presumably thermonuclear devices) to shake the ground over large areas, thereby triggering strain-relieving movements in the crustal rock. In the second, strain energy could be released by pumping in water which, acting as a lubricant, might conceivably cause adjacent blocks of rock to slip. The latter was inadvertently proved viable a few years ago near Denver, Colorado, when liquid chemical wastes were pumped underground as a means of disposal. The seismic tremors that clearly resulted were only slight and did virtually no damage, but the highly significant fact is that they *did* occur.

Use of either or both of these methods in the vicinity of the notorious San Andreas fault in western California could have fearful repercussions by bringing about the dreaded major earthquake already feared by the citizens of San Francisco and Los Angeles. This fault runs up the coastline of California and is probably the best known and assuredly one of the most dangerous. It is, in fact, a whole series of faults trending northwestward and represents a collision point between two tectonic plates. That to the west is moving northwest in relation to the rest of the American continent. Along parts of its course the San Andreas fault is able to achieve gradual slip. Strain is thereby continually released. Failure to achieve this means steadily accumulating strain until a colossal rip results in a massive earthquake. In certain regions, notably those around the cities and conurbations of San Francisco and Los Angeles, no such slip is taking place. In the case of San Francisco no slip has taken place since the great earthquake of 1906. These areas are regarded as "seismically quiet," and because of this the inhabitants may easily be lulled into a sense of false security. Here the two tectonic plates are locked. They might, in fact, be compared to the interlocking sections of a monstrous jigsaw puzzle. Though no movement is occurring, the strain is steadily building. One day, perhaps in the not too distant future, it will prove too much and something will have to give. This is precisely what happened at 5:12 A.M. on the morning of April 18, 1906, when the crust just north of San Francisco finally snapped. The resulting slip was extensive and comprised 274 miles of the fault from San Juan to Upper Mactoe. The maximum movement along the affected length of the fault was 21 feet. The rest is history.

The total movement along the fault during the last few millions of years appears to have been several tens of miles. In 1857 another slip occurred on the southern segment of the fault from San Bernardino northward. In 1971, movement on a small fault a short distance from the main San Andreas fault produced an earthquake in the San Fernando region. By the standard of California earthquakes this was a rather puny affair, though the damage was by no means inconsiderable. Virtually the whole of the Californian seaboard is highly sensitive, and clearly it would not take much, as conditions are today, to produce a very violent earthquake.

The energy released as seismic waves in the earthquakes of various intensities is colossal. During the time a major earthquake is being generated, the maximum power produced can amount to 3 million, million KW (3×10^{12} kw). The average seismic energy released per year is 300,000 million KW hrs (3×10^{11} kwh) and is derived principally from a small number of major shocks.

It must be obvious that here is a tremendous potential force concentrated at a few specific regions on the Earth; one always ready to go somewhere in the most devastating manner. In an area such as a locked portion of the San Andreas fault a violent earthquake would very likely be triggered by the underground detonation of a thermonuclear device.

So far as alien use of such destructive potential is concerned, we can only speculate. Would their knowledge of the unique geology of certain areas of our planet be sufficient? If it were, they might direct thermonuclear devices at vulnerable regions, devices perhaps capable of boring well into the crust of the Earth before detonating in a manner roughly analogous to that in which an armor-piercing shell pierces the armor of a tank or warship before exploding.

Like earthquakes, volcanoes are restricted, for similar tectonic reasons, to specific parts of the Earth's surface. The idea of deliberately engineered earthquakes inevitably suggests a similar technique with respect to volcanoes. Such action would perhaps be less easy to achieve. A locked and highly strained geological fault merely awaits the tectonic equivalent of a slight nudge. Volcanoes are rather a different proposition and it is difficult to see how they could be persuaded to erupt at will. On leaving a period of dormancy the eruptive process might be accelerated by the detonation of a thermonuclear device near the magma chamber or central conduit of a volcano. For obvious reasons, built-up areas generally steer clear of volcanic mountains, so that damage would not be comparable with that of a large scale earthquake, which spreads its effects over a fairly considerable area. External influences could certainly cause many volcanoes to become violently eruptive at once. Such influences would, however, have to be on the grand scale. Almost certainly the close passage to Earth of another planet of large dimensions would achieve this. In a highly volcanic area such as Indonesia, probably as many as two hundred volcanoes would salute the coming of such a heavenly wanderer. Something along these lines does not seem very practical as a deliberate action, however.

Use of Earth's magma might be better achieved by piercing the Earth's crust at a number of points, thus releasing vast and violent upwellings of magma from the mantle that lies beneath the crust. The crust of our planet, so far as continental regions are concerned, is relatively thick—about 35 kilometers, and more under mountainous regions. The crust under the ocean floor, however, is much thinner, amounting to only about 5 kilometers. So far man has not succeeded in boring down through the crust to reach the mantle, though an attempt was made (via the ocean bed) a few years ago under the code name "Operation Mohole." The term *mohole* is derived from the Mohorovicic discontinuity, named after its discoverer. This lies at a depth of between 5 and 40 kilometers beneath the Earth's surface and marks the lower boundary of the Earth's crust. For a variety of reasons this project was unsuccessful and was probably a waste of time, resources, and money from the very outset.

Back now to our alien attackers, who by virtue of a tremendously advanced technology are able, while orbiting Earth, to direct high velocity, spinning projectiles capable of boring their way through the crustal rocks beneath the seabed. When they reach the junction between crust and mantle, their thermonuclear warheads detonate. Up the fractured crust spews a flood of seething red-hot magma which at once comes into violently explosive contact with seawater. In shallow waters only a few miles from highly industrialized, densely populated coastal zones, the horrifying, escalating nature of such an assault leaves little to the imagination.

Probably the most exotic form of alien warfare (though already being envisaged by scientists on Earth) involves tampering with the electrical behavior of the ionosphere, the ionized region of the atmosphere extending from about 50 kilometers to some hundreds of kilometers above our planet's surface. By punching "windows" in the ionosphere using nuclear devices, radio communication could be greatly disturbed, a vital factor if we were already under attack. Even worse, such artificial "windows" in the ionosphere could prove lethal by permitting the ingress of very short wavelength ultraviolet radiation. Such radiation is known to damage biological systems, causing skin cancers in humans and genetic damage to crops.

From this point it is only a short step to a postulated technique with connotations of a really old favorite in the realms of science fiction—the taking over of men's minds. Nowadays there rarely seems to be a single television science fiction series in which at least one episode does not involve the taking over of men's minds by aliens. Somehow it always seems highly improbable, despite contemporary advances in the field of hypnosis. It cannot be denied, of course, that if aliens requiring our world could somehow transform us into a race of docile, obedient "zombies," their task would be rendered very much simpler. No violent action would be required, no large-scale destruction. Just a simple, smooth takeover with the aliens able to do with us afterwards as they wished. (A power that no doubt many of the world's politicians wish they already possessed.)

It is now being suggested that the natural waveguide between the ionosphere and the surface of the Earth could be used to propagate very low frequency radiation through it in such a way as to affect the electrical behavior of an individual's brain activity. The activity of the human brain is characterized by certain rhythms having a frequency in the region of 5 cycles per second (5Hz). This is quite close to the lowest resonant frequency of the ionospheric waveguide, 8 cycles per second (8Hz).

At the present time very little is known about the efforts of weak oscillating fields on human behavior and response. This is clearly a field for investigation and one that could offer much in the way of interesting results. A few experiments have already been carried out in which human beings were exposed for about fifteen minutes to low-strength fields (about a few hundredths of a volt per centimeter). The result, in most cases, was a small but measurable deterioration in the person's performance. The fields employed, however, were a thousand times *greater* than the observed natural oscillations in the Earth's waveguide. The concluding comments regarding these experiments reveal the possible future potential and are such as to send a few shivers down the spine. "If methods could be devised to produce greater field strengths of such low frequency oscillations, either by natural (e.g., lightning) or artificial means then it might become hypothetically possible to impair the performance of a large group of people in selected regions over extended periods."

Just after the above words had been written, the attention of the author was drawn to an article appearing in the December 15, 1977, issue of the British periodical *New Scientist,* much of which is relevant to this theme. The article refers to an alleged new Soviet weapon that according to the writer, embodies a low-frequency radiation beam, which the Russians are said to have been testing for the past couple of years.

The weapon, as described, is allegedly based on the work of Nicola Tesla, whose name has already been given to the unit of magnetic flux (the tesla). Tesla at one time became greatly preoccupied (as did a number of his contemporaries) with the effective transmission of electrical power without the use of wires. His argument centered on the fact that Earth could be used as a conductor and would be responsive as a tuning fork to electrical vibrations of a particular pitch. Becoming more specific, he claimed that transmissions of sufficient energy at low frequencies (6–8Hz) would travel through the Earth and behave as terrestrial standing waves. To prove his point, during 1899 he constructed in Colorado Springs the largest induction coil ever seen and by its use succeeded in lighting nearly one hundred light bulbs twenty-six miles distant by transmission of electrical energy through the ground *without* wires.

At first, Marconi shared Tesla's belief in such possibilities, but soon changed to air as the medium. His work in this sphere culminated in the first radio transmission across the Atlantic in 1901. Tesla was more persistent and continued in his endeavors to develop a scheme by which a signal close to the fundamental resonant frequency (8Hz) would, he reckoned, travel through the Earth and be reflected back from the other side in the form of a standing wave. From all accounts, Tesla appears to have been quite a character, announcing on his seventy-fourth birthday his alleged invention of a so-called "death beam" capable of clawing aircraft out of the sky at a range of 250 miles. Tesla never furnished any detailed description of this device, however.

The alleged Soviet weapons system referred to earlier is supposedly intended to control people and is allegedly based on Tesla's ideas. Five transmitters are already reputed to have been built, one near Riga and the other in the vicinity of Gomel, both in the western part of the USSR. Again, according to accounts, the weapon has one modification not conceived by Tesla. In 1957, W. O. Schumann discovered a different resonance—in the cavity formed by the Earth's surface and the lower layer of the ionosphere. This resonance finds use in submarine communications systems. The fundamental frequency is 8Hz, which is very close indeed to that quoted by Tesla. It would seem, therefore, that as well as being transmitted through the Earth the Tesla beam must be transmitted around it.

The frequencies in question are reputedly close to those of the human brain and so, in a way somewhat analogous to radio jamming, they could be used to affect the brain. It is postulated that various pulse rates could produce effects ranging from drowsiness to aggressive tendencies.

Already some odd claims have been made regarding the results of Soviet tests with this supposed device. It might be just as well, however, to accept them with caution. For what they are worth they include a "humming in the ears experienced by certain residents in Britain" (as one of these residents, this writer has so far remained unaffected), violent riots in parts of the world (though these can usually be generated without any assistance from Tesla beams) and considerable climatic disturbance in Canada (the precise nature of which remains unstated). A further claim that the recent Peking earthquake was caused by such excitation of a particular point on the Earth's surface thousands of miles distant should hardly improve Sino-Soviet relations.

To excite the Earth at the fundamental frequency would

require fantastic power and a positively enormous antenna system. The latter would have to be something like a buried copper plate over an area of 20 square kilometers. Moreover, it is felt that construction and testing of such a device would have been detected easily by means of satellites. Whatever the potential of Soviet very low frequency transmissions, they almost certainly do not possess the power attributed to them.

The position becomes a little bizarre in the end in view of reports (apparently made in good faith, though for obvious reasons likely to be received with a certain scepticism) that Tesla was "contacted" several times by "extraterrestrials" who imparted their knowledge to him, enabling him to produce detailed drawings of devices without having previously worked on them. It would be most convenient if aliens were considerate enough to let us know of their weapons systems in advance. Whatever might be said about Tesla's theories and any subsequent development of them by the Soviets or anyone else, this latter embellishment seems somewhat improbable.

Impairing people's performance is hardly the same as taking control of their minds, thereby manipulating their thoughts and actions. Nevertheless, the idea that we ourselves on Earth are already taking the first steps along this strange and frightening road is revealing. It is a concept that opens up all sorts of alarming possibilities, including that of totalitarian rule. In the circumstances, it is inevitable that we must wonder what degree of capability civilizations millennia in advance of our own have already achieved in this respect. Dare we consider that certain radiations emanating from alien starships could one day transform us into obedient robotlike creatures, prepared at any cost to do the work of the aliens for them? That would indeed represent a fifth column to end all fifth columns!

12. AND DARKNESS LAY OVER THE EARTH

In almost any form of conflict it is reasonably safe to predict that if one side could be rendered temporarily unseeing, victory would quickly and fully accrue to the other. This is something we have already touched upon briefly in the preceding chapter in connection with thick fog. The nearest we have ever come to this in the course of the many wars and battles here on Earth has been the use of smoke or dense chemical vapors either to screen advancing forces from the enemy's view or to hide, as far as is possible, evidence of buildups for coming offensives.

As a technique it has never been particularly successful, relying as it does on wind direction and strength. In fact, rather like gas warfare, it can be something of a two-edged weapon should the wind change. An advancing formation of troops or tanks feels very exposed, almost naked, if a sudden change in the direction of the wind causes the protective screen of thick vapor to be blown away.

It is no doubt feasible that a highly technological alien society would have given this question considerable thought. It is equally feasible that they could have developed techniques as a result of which their adversaries were unable to see what was going on until it was too late. Such crude terrestrial weapons as mustard gas, which can blind and blister as well as asphyxiate, are not being considered in this context. These weapons are only effective over relatively small (though sometimes vital) areas and can, in the end, prove as big a nuisance and embarrassment to

114

the user as to the victim—physically as well as morally. What is really envisaged here is the laying down of some dark cloud, the introduction of a realm of dense black darkness which, even if it could be pierced by artificial illumination, would still represent an enormous handicap.

Oddly enough this is not a scheme that, so far as this writer's researches have revealed, has commended itself to the writers of science fiction. One story from over forty years ago did latch on to the idea, though not in the context of a deliberate black cloud used in a military role. In this instance there lay squarely in Earth's path as it rolled through space a great dense cloud of intensely black gas. Just how it came to be there was not explained. It is perhaps just a little unfortunate that the author of this entertaining and novel tale decided to be so specific regarding the *nature* of the gas. Certainly his choice was an unfortunate one and indicated rather clearly that he had not done his chemical homework. The gas, he declared, was argon. Now argon is not, never was, and never will be, a dense black vapor. It is, in fact, one of the so-called inert gases (the others in the family are helium, neon, xenon, and krypton). They are all present in exceedingly minute proportions in our atmosphere and will only show the most limited reactivity under very special conditions. This lack of affinity for the other chemical elements is due to the electronic configuration of their respective atoms. They cannot, of course, be breathed, so that an enveloping cloud of argon stealing over our planet would only further dilute the oxygen we breathe. When this black cloud of "argon" finally shrouded Earth in its entirety, it was as if all mankind had been rendered totally blind. Normal tungsten-filament electric bulbs, matches, and flames, simply could not pierce the frightening stygian darkness. It was soon discovered that the darkness could only be penetrated by the use of sodium vapor lighting—the typically orange lights found in so many of our cities and in highway overhead lighting. This has some advantage during mist and fog (which is probably why the author hit upon the idea). Dense black argon that can be illuminated only by sodium vapor lighting certainly represents a peculiar reorientation of fundamental chemical principles. Still, it is the basic idea that is so interesting in our context, not the nature of the gas.

The concept of our planet running into a vast pocket of some material is perhaps not quite as ridiculous as it sounds, though it is unlikely to be argon or have such peculiar physical properties. A recent study by R. J. Talbot of Rice University, and M. J. Newman of Cal-Tech, provides some careful estimates of the frequency and consequences of encounters between the Solar System and dense interstellar dust clouds. Using the best available statistics relating to the properties of such clouds within the galaxy, the authors conclude that during its lifetime (about 5,000 million years) the Sun (and Solar System) has possibly passed through many such clouds—135 having a density corresponding to one hydrogen atom per 100 ccs and 16 with at least ten times that density. Clouds of this nature are thought, in the main, to lie in the spiral arms of the galaxy.

Encounters of this type would expose the Earth to the full galactic cosmic ray flux. Our planet would also accumulate helium and many of the heavier elements as well as very large amounts of dust. The most important effect, so far as our world is concerned, would be a pronounced increase in the intensity of the shortwave radiation to which it is exposed. This would have drastic effects on our climate and would also present a considerable hazard to life. It is unlikely that the large amounts of interstellar dust coming into our atmosphere would have the effect of causing a worldwide blackout, however.

These are interesting digressions, but it is time we returned to our world and its temporary shroud of thick, black "argon." Consider the circumstances. The entire planet Earth is enveloped in a darkness that effectively and entirely cuts off the light from Sun, Moon, and stars, an intense jet blackness that flame and normal lighting will not penetrate. Almost in an instant mankind is helpless. Aircraft circle airfields helplessly until their fuel runs out or they collide with others; trains overshoot the stops at terminal railroad depots and pile up with devastating results; ocean liners sail blind and blunder catastrophically into piers as they try to make port. Roads and freeways are littered with burning heaps of crashed automobiles—though only the heat of their cremation is apparent, not the light. In this instance, presumably because a readable story had to be made out of life amid the stygian blackness, the author introduced the idea of illumination by sodium vapor lighting. Suppose, however, that even this were not possible. Civilization would begin to crumble. Only those already blind for many years would be in any sort of position to keep going. To them nothing would have changed. On their shoulders would fall the awful responsibility of somehow keeping civilization ticking.

Naturally, the reader will be inclined to comment, quite justifiably, that such a gas or dense vapor used in a military sense would be as great a handicap to the attackers as to the defenders. This is a good and valid point. We must assume, however, that those employing such a weapon would also have first developed some means whereby *they* at least could see through it. Without this precondition, use of the gas would be utterly nonsensical.

It goes without saying that our knowledge does not extend to any material with the characteristics we have just described. Indeed, physically and chemically, such a gas or vapor could represent an impossibility. As always, of course, we must be wary in making statements of this sort. It is impossible for us to predict the chemical and physical expertise of races millennia ahead of our own. We tend to feel that our grasp of such matters is both extensive and considerable. And it is—compared to that of a hundred years ago. But how will our descendants a century or so from now view the chemical and physical expertise of the late twentieth century? Not, perhaps, as anything very spectacular. We must therefore be careful before saying that a particular thing is impossible. Some things may well be impossible by anyone for all time. Others may just *seem* impossible to *us* at the *present* time.

Let us assume then, for the time being, that an alien force from the stars, bent on taking over our planet, has the power to cast a shroud of total and impenetrable darkness over the Earth. They, because of a lens made from some special material, retain the power to see through the gas. The takeover could be totally effective in the very minimum of time and with little or no bloodshed. By the time the black cloud had been rolled away and we stood blinking uncertainly in the strong sunlight, the aliens would be in complete and absolute control. It would then simply be a matter of obey—or else! The circumstances might suggest an alien attacker with an essentially merciful nature. It could equally well involve one desirous of securing our planet, its resources, its facilities, and a ready-made slave force with the minimum of fuss, time, and damage.

The mechanics of cocooning, however temporarily, an entire planet in a thick black shroud (assuming such a material is possible) would hardly represent a simple undertaking. We must envisage a fleet of alien craft descending into the upper layers of our atmosphere, each one pouring forth great, streaming gouts of dense, black, cloying vapor. It would be perhaps like a colony of squids on a gigantic scale. These marine creatures (a branch of the cephalopod family) eject a stream of the dark brown pigment sepia, which spreads quickly as a rolling dark, obscuring mass over the water behind them. A sepialike material with aerosol properties would probably have been a better choice than argon in the story just mentioned. Perhaps this material could provide us with the first, faint glimmerings of techniques already perfected by advanced beings elsewhere in the galaxy. To be effective such a vapor would have to be resistant to dispersion—which would seem to necessitate a structure in which the individual molecules clung to their neighbors with considerable tenacity in three dimensions. This would simply be what chemists term a high polymer. Such compounds naturally have a high molecular weight. What we are envisaging would quickly sink to the ground and that would be that. A polymer hardly seems the answer. Moreover, though excessive dispersion would render the material equally useless, a certain element of dispersion would be necessary if the material were to spread itself uniformly throughout the atmosphere. In effect, what would really be aimed at is some form of "emulsion." Chemically, the term emulsion means a dispersion of oil globules in water or some other liquid in which oil is not soluble (or vice versa). Materials known as dispersing agents break the oil up into a myriad of tiny bubbles which spread themselves uniformly throughout the water. So long as the dispersing agent remains effective, the emulsion remains stable. Should the dispersing agent for some reason or other become noneffective, the oil and the water separate out in the typical fashion of nonimmersible liquids, the oil settling out in the top layer. A form of "emulsion" can also be formed out of solid particles in air. Shortly we shall be mentioning a volcanic phenomenon known as a *nuée ardente,* or glowing cloud. This is a mass of very hot particles of pumice and glass shards spead throughout a volume of superheated volcanic gases. By now we can see, though still only hazily, the kind of black obscuring mass our alien invaders might employ. The material, whatever its nature, would have to disperse itself uniformly among the molecules of oxygen and nitrogen that constitute our terrestrial atmosphere. Two conditions would have to be satisfied. The first is that this "emulsion" would have to remain stable long enough for the takeover of our planet to be completed. The second is that the material would have to be nontoxic to us poor terrestrials, since it would be passing into our lungs. If, on the other hand, the aliens wished simply to dispose of Earth's indigenous inhabitants, it could be rendered toxic. The aliens themselves, with appropriate respirators, would be immune. A few chapters back we discussed some of the ways in which alien invaders could rid themselves of the nuisance of a terrestrial population. Here is another one, almost diabolical in its effective simplicity.

This highly lethal extension of the theme is a distinct digression from the idea of synthetic darkness and merely represents chemical warfare on the ultimate scale. A gas, vapor, or dispersion designed simply to asphyxiate does not require obscuring qualities as well. The aliens merely lay down their "superinsecticide," which could be colorless, and all mankind dies. Perhaps if merciful, they would only put us to sleep while they went about their business. It would certainly be a changed Earth and social order when we finally woke up. Depending on the aliens it might even represent an improvement.

We cannot wax too moral on this theme after using chlorine, phosgene, lewisite, and mustard gas during World War I. These war gases and their logical descendants were never used in any theater of World War II. We suspect this had nothing to do with moral issues. It is much more probable that both sides were equally apprehensive about payment in kind. The "black gas," used freely by the Martians in H. G. Wells' *War of the Worlds* was of the asphyxiating kind, highly effective over large areas; so we see that this concept was envisaged at the very dawn of twentieth century science fiction writing. Within two decades it had come to fruition on the blood-soaked fields of Flanders. That the theme of asphyxiating vapors has not been used more often in science fiction is probably due to this fact, which at once tends to render it less spectacular and rather old-fashioned, especially when compared to death rays, disintegrating beams, and other ultrasophisticated methods of dealing out death and destruction.

The use of a black, planet-enveloping cocoon against us seems much more feasible and realistic if we envisage a material less total in its effect. Even if our present methods of illumination during the hours of darkness were effective, we would still be at a very considerable disadvantage for we would, in effect, be fighting an invader having superior weapons during a perpetual night. And that invader, because of some other sophisticated technique he had devised, would not be so inhibited.

Obscuring the light from the Sun may seem a highly fanciful prospect, and certainly so far as we are concerned at the present time this is so. A little reflection, however, ought to tell us that cloud formations are always doing this over part of our planet. At least the Sun itself becomes obscured even though its light is only dimmed. Very little of its light filters through the dark gray base of cumulonimbus (thunderhead) clouds, however. The appraoch of an electrical storm is often heralded by such expressions as "the sky is as black as ink." It is never quite that black, but at least this provides an idea of how light-obscuring these clouds are. From time to time the light of the Sun may be very effectively obscured over portions of the Earth by virtue of the vast dust clouds pouring upwards from volcanoes in large scale eruption. In 1883, Krakatoa, in the Straits of Sunda between the islands of Java and Sumatra, achieved this. It achieved a great deal more in fact, for it also sent some of that same dust on a world-girdling flight, which led to fantastically colored sunsets and sunrises many thousands of miles from the scene of the eruption. A genuine "blue" Moon seen by the writer in Scotland in the early 1950s was also attributed to a volcanic eruption many thousands of miles away. (Later that evening the blue was replaced by pink.) All this emphasizes the fact that haphazard natural phenomena can dim or modify the light of the Sun over considerable portions of our planet's surface. To effect this much more totally with the resources of a highly advanced technology might not therefore be beyond the bounds of practicability. And even if the natural tendency of our world's wind system is to dissipate smokes, vapors, and dust clouds, our alien tormentors would no doubt remain on station to replenish the supply so long as this was necessary.

Anyone who has witnessed the tremendous ash and dust output of a major volcanic eruption will appreciate how fantastically thick and dense these can be. Two notable examples of this were Bezymianny in the Soviet Union in March 1956, and Hekla in Iceland in March 1947. Equally dense are the terrible *nuée ardente* or "glowing cloud" eruptions from volcanoes such as Mont Pelée in the island of Martinique in the West Indies. In fact, the extent to which volcanic dust and ash being hurled into the sky will obscure light is well seen in the memorable account by Pliny, the Younger, of the famous eruption of Vesuvius, which destroyed the Roman towns of Pompeii and Herculaneum in 79 A.D. It includes these very relevant lines, "Elsewhere the dawn had come but here it was night, the blackest and thickest of nights." Later, during his escape from the stricken area he wrote: "We were enveloped in night, not a moonless night or one dimmed by cloud, but the darkness of a sealed room without lights. Only the shrill cries of women, the wailing of children and the shouting of men were to be heard!"

A few lines back we mentioned the terrible volcanic phenomena known as *nuées ardentes* or "glowing clouds," which are a typical feature of certain types of explosive vulcanism. Though this represents a further digression, here also might be the basis for an alien weapon of a very horrible kind.

One of the earliest recorded instances of a *nuée ardente* undoubtedly occurred during the catastrophic eruption of Mont

Pelée, a small and rather undistinguished volcanic peak some two or three miles or so from the thriving city of Saint Pierre on the Lesser Antilles island of Martinique. At 7:59 A.M. on May 8, 1902 (Ascension Day) one of these glowing clouds suddenly burst from a ruptured flank of the volcano. In the brief space of three minutes it had rolled across the intervening terrain. At 8:02 A.M. it reached Saint Pierre. At that precise moment the city died. When it had passed, almost the entire population of 30,000 souls had perished—and in the most excruciatingly painful and horrible way. There were but two survivors, one of them a condemned murderer awaiting execution in the death cell of the local prison—surely the crowning irony.

A *nuée ardente* is simply a glowing cloud of superheated gases and incandescent solid particles. Being heavier than air, it hugs the ground while rolling over it at speeds of up to one hundred miles per hour. So great is the heat contained in one of these glowing clouds that trees and telegraph poles in Saint Pierre were not just charred on the outside. They were in fact carbonized to their very centers. All this occurred in the space of two or three minutes. The city was itself superheated when the glowing cloud enveloped it. When the latter moved on toward the sea, every building ignited spontaneously as the oxygen returned. Temperatures, it is reckoned, were between 600°C and 1,000°C. On reaching the sea the water boiled almost at once and ships lying in the harbor became flaming infernos within minutes. Dying hands tried desperately to take dying ships out to sea and safety.

The basic essentials of a *nuée ardente* are therefore gas, a dispersion of minute solid particles, and the velocity and heat to provide its devastating punch. As a possible alien weapon we can visualize containers in which gas and an emulsion of solid particles are highly compressed. A heat source, perhaps thermonuclear, raises the mass to tremendous temperatures. It is then ejected under colossal pressure by a directing orifice or jet. Perhaps no single one could equal the might of a natural volcanic glowing cloud, but several working in unison could almost certainly devastate a city or reduce a defense line. A heavier-than-air emulsion of glowing solid particles in superheated gas rushing along the ground with a velocity of perhaps one hundred miles per hour is a highly daunting prospect. Battlefield heroics would be of little or no avail. To stand and fight would be only to be carbonized and fall. Photographs are available of a few of the unfortunate victims of the May 1902 holocaust at Saint Pierre. They are not a pretty sight—disemboweled, semi-carbonized shapes that only minutes before had been active, healthy men, women, and children going about their everyday routines.

In a sense such a weapon could be seen as a gross extrapolation of the flame throwers and napalm devices that can be employed so devastatingly against machine-gun nests and pill boxes on present day terrestrial battlefields. It would be hotter, more explosive, and effective over greater distances and areas. Sophisticated emulation of Vulcan's forge might not prove the ultimate in offensive weaponry. But neither would it hand out a merciful death to those whom it was directed against.

Earlier in this chapter we confined ourselves to the idea of a nonlethal obscuring cloud projected into our atmosphere by the advance guard of an alien space fleet. Clearly this concept could also apply in space itself during a battle between rival fleets. Total chaos would inevitably result if one side possessed such a weapon and were able to envelop its opponent's fleet in it. Ships of the latter would be "blind," indeed unable to see each other, unable to see their adversaries, with whom they would probably collide catastrophically. Those not so destroyed could be picked off mercilessly and cold-bloodedly by their opponents who, before developing such a weapon, would also have devised some method to see through it themselves. And if it is thought that radar techniques would still enable the optically "blind" ships to see their way and their adversaries, we would also suggest electronic jamming by their opponents to remove this possibility.

In closing this chapter we should briefly mention an idea that has been used and developed in one or two science fiction tales—that of temporary blindness for Earth's inhabitants, produced not by opaque chemical clouds, but by some appropriate radiation capable of acting adversely upon the optic nerves or retina of the eye. How this could be directed against the entire population of our planet, or of any other inhabited planet, was never made particularly clear by the respective authors concerned— perhaps for the reason that it was too difficult. A probable technique might involve the putting into orbit around our world of a kind of minuscule "sun" piercing the clouds and drenching the Earth with sight-destroying radiation.

In the general context of this chapter, an interesting idea was conceived in 1974 by Fred Hoyle, one of Britain's leading cosmologists, in a science fiction novel. This involved an alien race from beyond the Solar System who, having been defeated in its first attack upon Earth, decides to destroy all life on this planet by launching a large and dense cloud of hydrogen gas which must envelop our world and permeate its atmosphere. The aim on this occasion is not one of asphyxiation. It is far more diabolical since the hydrogen, having once diffused and thoroughly mixed with the abundant oxygen in our atmosphere, would render that atmosphere a veritable time-bomb of gargantuan proportions. Hydrogen mixed with oxygen represents a highly combustible and explosive mixture which, if deliberately or accidentally ignited, would at once and with great violence transform our atmosphere into the very ultimate in raging fire-storms thus quickly transforming Earth into a scorched and lifeless cosmic cinder.

13. THE BATTLE OF THE BEAMS

No volume dealing with future interplanetary or interstellar conflict could be considered complete without a discussion of beam warfare. We have touched on beams to some extent already, but now it is time to consider the subject of melting or disintegrating beams at greater length. These, of course, are essentially different from the pure heat rays we have already discussed.

The first candidate in this field must be the laser, which would almost certainly play a major role in any future indigenous large-scale conflict here on Earth. Equally likely is its utilization as a dominant weapon by advanced galactic communities.

So far as this planet is concerned, the development of the laser has been surprisingly rapid and it may already be considered to have outclassed radio in this respect. Indeed, there are those who feel that among certain alien civilizations the coming of the laser might easily have preceded that of radio. On this we can only speculate and, although this writer believes it unlikely for a number of reasons, the possibility does remain.

When practical laser beams were first introduced they had, not surprisingly perhaps, an almost science fiction aura about them. Now, only two decades later, they are accepted as a normal part of the modern technological scene. In a sense this epitomizes the atmosphere that has been engendered by the recent rapid acceleration in technological progress. There is, for better or for worse, an increasing readiness to take all new inventions for granted, irrespective of how far-reaching in their implications they may be. For this state of affairs, there can be little doubt that the news media is largely responsible. Much of the information they disseminate on technical matters is so popularized that it winds up as hazy half-truths. As a consequence millions of people latch on to popular words thereby deriving half-baked notions. In time this is something we could come to regret—if we are not already doing so.

First, a few brief notes on the laser and its inception and development on this planet are in order. The term *laser* is derived from the initial letters of the words "*l*ight *a*mplification by *s*timulated *e*mission of *r*adiation." In general terms it is a device that produces the purest and most intense light known. In fact, its light has been compared to that emitted at the *surface of the sun.*

A normal beam of light, from a searchlight, for example, possesses a considerable element of divergence or spread. The rays of light coming from the precise center of the beam are at right angles to the source, but from all other points the angle is greater than 90°. This means that the rays are spreading out, and therefore the beam, which is merely a combination of all the rays, is doing precisely the same thing. The effect with respect to the initial portion of the beam, i.e., in the immediate proximity of the source, is slight and of little or no consequence. The farther the beam travels, however, the greater becomes the degree of divergence. Thus, a seemingly brilliant and tight searchlight beam would illuminate a circular patch of cloud, say at 8,000 feet, of much greater radius than that of the beam leaving the searchlight. In the detection of raiding enemy aircraft during World War II this feature was obviously not without merit, although the illuminated area was much less bright.

Now if all the individual rays comprising the beam could be rendered *parallel* to those emanating from the center of the projector, a tight and sharply defined beam would result. This is precisely what the laser beam achieves. Divergence or spread simply does not occur.

Normal white light comprises components at a variety of

different wavelengths, and these waves travel in virtually *every direction.* Such light is termed *incoherent light.* The light produced by a laser, however, is virtually of *one wavelength* only. As a consequence, the waves are *unidirectional* and effectively reinforce one another. The resultant beam is therefore straight, narrow, and very sharply defined over considerable distances and is an example of *coherent light.*

The laser story can be said to have begun in 1958 when two American physicists, R. N. Schwartz and H. H. Townes of the Bell Laboratories, published a paper in which they outlined the theoretical basis for such a beam. Practical application of the idea was later demonstrated by another American physicist, Dr. Theodore Maiman of the Hughes Aircraft Company.

As with so many technical matters, the underlying principle of the laser is comparatively simple and straightforward. At the center of the device is a thin rod of synthetic ruby. (Natural ruby is a red transparent variety of corundum or alumina [Al_2O_3]. It can be duplicated synthetically from purified ammonium alum and chromium sulphate.) The ends of the rod of synthetic ruby are so treated that one functions as a mirror giving total reflection and the other as a partially transparent surface. The ruby rod is surrounded by a flashtube rather similar to the kind used in high-speed photography. When this flashtube is discharged a very intense light is immediately generated. The effect of this is to excite the electrons within the ruby rod. Though chemically a form of aluminum oxide, ruby is distinguished by the fact that some of the aluminum atoms within its crystal lattice have been replaced by those of chromium. These chromium atoms, unlike those of aluminum, possess electrons that are not "locked" within the crystal lattice. Being mobile they can be raised to a higher energy state, and it is precisely this function that is achieved by the discharge of the flashtube. As the excited electrons revert to their normal energy level or unexcited state, they emit photons of light energy. These traverse the entire length of the ruby rod, rebounding between the two mirrors. By so doing they cause other excited electrons to emit light also. The final and cumulative effect is a virtual torrent of red light oscillating several million times between the mirrors within a few millionths of a second. Ultimately, the light is rendered so intense that it *pierces* the partially transparent mirror to emerge as a laser pulse of coherent light.

The original laser beam was not capable of continuous operation, in that it produced only *pulses* of laser light and not a continuous steady beam. It was not long before this deficiency was corrected and in 1961 a continuous laser beam apparatus was perfected. In this the ruby rod was replaced by a helium-neon gas mixture. It is an established fact that certain other materials can be used in place of ruby. Gallium arsenide, for example, permits the production of a laser beam capable of functioning at low temperatures. Carbon dioxide, which can also be used in this role, leads to the emission of high-temperature infrared light.

The advent of the laser beam at once brought the old science fiction nightmare of lethal beam warfare closer. Even the intense red color was highly appropriate. Something that had hitherto been regarded as entirely the figment of imagination now seemed perfectly feasible, given sufficient time for the necessary degree of research and development. Early practical experiments did much to reinforce this idea. During May of 1962, scientists at the Massachusetts Institute of Technology used a pulse ruby laser to "illuminate" a 20-mile diameter portion of the moon's surface. The resultant reflection from the moon was detected quite conclusively by ultrasensitive measuring equipment. The advantage of a laser beam over a beam of conventional light was illustrated by the fact that a beam from the latter would, because of spread, have "illuminated" an area equal, not to twenty miles, but to one several times the diameter of the moon which is 2,160 miles. By 1962, the far side of the moon had been crudely photographed and certain unmanned space vehicles had been deliberately crashed on its surface. The feat with a laser beam may not have matched these more spectacular ones, but in its implications for the future it may well have exceeded them.

Since making its debut in 1962, the laser has been used for increasingly diverse scientific and technical applications, including eye surgery, isotope separation, and optical alignment. For many applications it is now desirable to have pulsed laser beams. This, of course, is something of a throwback to the early days of lasers, which started out as pulsed beams and were subsequently modified to give continuous beams. The durations of the pulses called for today, however, are much shorter than those of the original pulses. It is now possible by virtue of a process known as mode-locking to generate pulses having a duration of less than a picosecond (a billionth of a second, taking a billion as 10^9). Such ultrashort pulses have very widespread uses, e.g., high-speed photography and optical communications systems.

By focusing the output of a mode-locked laser, very high intensity beams can be produced with relative ease. These range around *10^{15} watts per square centimeter* (10 thousand billion) and even higher intensities are possible if the pulse is first amplified in an auxiliary laser system. The potential of such a device may be better appreciated by comparing it with the value for average, unfocused intensity of sunlight on the Earth's surface, which is only about *one tenth of a watt* per square centimeter.

The tremendously high intensities obtainable from picosecond laser pulses have many applications. They hold considerable promise in the realm of nuclear fusion. We have already considered nuclear fusion in the context of interplanetary Armageddon. It is believed that if amplified laser pulses are focused on a pellet of an appropriate compound, such as lithium deuteride (a compound of lithium and deuterium), the compression they might create could be sufficiently great to initiate a fusion reaction. The impact of one of these pulses can instantly create local pressures as high as a *million* atmospheres. Such pressures, whatever their future potential in initiating thermonuclear reactions, are already highly effective in punching holes in solid material—without affecting the surrounding material.

So there we have it! The laser's existence began to all intents and purposes in 1958. Today, only two decades later, we have developed it into something with the power of a million atmospheres, able to pierce holes quickly and effectively through solid materials. We are also able to envisage it as a key perhaps to the production of vast and limitless energy supplies by thermonuclear fusion—all this in twenty years by a civilization that could be, and very likely is, several millennia behind that of other galactic peoples. To what extent then have *they* developed this amazing device?

In the field of weaponry it does not take very much imagination to comprehend the possibilities inherent in a device able to punch holes through solids at a distance or to unleash, in a controlled way, the terrible tiger of nuclear fusion. Suddenly the spectacle of an alien starship boring a lethal hole through the metal skin of a defending space vehicle by means of a red beam is no longer science fiction. In 1939 it was the purest fantasy. In 1979 it is already on the verge of possibility. By 2039? But why think in terms of 2039? Other galactic beings will probably have had this power and very much more for several centuries. And someday they might come our way.

Having seen something of the power and potential of the laser let us now consider what other possible forms of beam weapons might have found their way into the armories of galaxy-roaming, planet-hungry aliens. A likely candidate must be the high-energy particle beam. Such a beam is already being considered here on Earth as a possible means of destroying incoming enemy missiles in any future conflict. So far, however, it is not possible to direct a beam of high-energy particles into the sky in the way a searchlight beam is directed.

Two different types of particles might serve as the "ammunition" for such a beam—perhaps by us in the future, or by *others* now! These are known as hadrons and leptons. The former are nonpenetrating, whereas the latter have a penetrating potential. It is unfortunate that the most easily accelerated particle, the familiar proton, is of the nonpenetrating kind. Proton beams directed by us at a low-orbiting hostile starship would suffer attenuation by a factor of some *hundred million* by the time they had passed through Earth's atmosphere. By then they would be scattered over a very wide area and be far too weak to cause any damage to the intruding vessel.

Muons (which result from the decay of mesons) are highly penetrating and can be directed in a beam. Unfortunately their initial intensity is far too low to be in any way damaging to the target. Probably the only way we could direct a truly damaging beam against an invading galactic vessel would be first to punch a "hole" through the thick, intervening layers of atmosphere and then to "fire" a proton beam up this hole. It is now believed that a rapid series of pulses from an accelerator might enable us to drill such a hole by intense heating of a column of air. The column is envisaged as reaching a temperature of 10,000°K with a diameter of only a few centimeters. This would, it is thought, create a reasonably clear path for the proton beam, enabling it to reach its target only slightly attenuated. As is so often the case, however, eliminating one difficulty only leads to or exposes another. The second problem in this instance is that the beam would very likely be rendered unstable by virtue of the now ionized state of the air following such rigorous thermal treatment.

The effect of this would probably be to make the proton beam whip about like a dropped high-pressure fire hose. In such circumstances it would obviously be difficult indeed to restrict the beam to the cleared atmospheric channel. Consequently its chances of impinging upon the target would be very greatly reduced at best. Nor is this the only additional problem. Even though the beam had successfully negotiated its narrow channel through the atmosphere and still remained capable of delivering the several megajoule punch necessary, there would still remain the vital question of accurate aiming. We must remember that the starship, being in orbit, is moving. Moreover, it could also be using evasion tactics. And there is the further problem of keeping up a sustained rate of fire. If the starship launched a host of shuttle craft or thermonuclear warheads at targets on the surface of our planet, the beam would have to be very adaptable indeed. In this respect a beam such as has already been suggested for use against missiles (before they are MIRVed) would probably be necessary. This postulated proton beam is assumed to be one of 5 GeV delivering 10 megajoules per burst with a diameter of 3 centimeters and a rate of fire of 10 pulses per second. Such a beam, it is reckoned, would very rapidly take aluminum to its melting point of 661°C. A starship could therefore be vulnerable to attack by such a weapon. Indeed it is not very hard to imagine the catastrophic decompression results that must inevitably ensue if a great hole were punched through the outer skin of a space vehicle. For those inside the vessel demise would be sudden, unpleasant, and probably rather messy. If aliens already possessed a weapon of this nature (and it is likely that they would) its effects, especially in this respect, would be well known and understood. As a consequence they would probably have the knowledge and expertise enabling them to counter the attack. This they might do by the use of alloys or metals impervious to such attack or through the employment of a device or technique whereby the lethal beam could be deflected. As many of us are aware, the U.S.S. *Enterprise* of "Star Trek" fame employs a shielding device that deflects the destructive rays of hostile craft. Indeed, it would seem to owe its continuing survival to this.

Thus far, we have only been thinking in terms of our using such a weapons system against an alien invader. We do not have such a system at the present time. Out there in the darkness there may be others who *do*. Therefore, all we have so far contemplated doing to *them* from the ground could be done to *us* from the sky. We would be highly vulnerable against a high-energy particle beam capable of melting aluminum and its alloys. Equally vulnerable would be our defending spaceships closing with the foe. Indeed, much of the invincibility that we mentioned a few chapters back with respect to a heat ray would be equally enjoyed by alien invaders equipped with this weapon. Fiction writers of three or four decades ago could freely envisage alien attack by a multiplicity of mysterious colored rays and then shrug off all ideas of real menace in the safe assumption that the things they had dreamed up were either impossible or lay so far ahead that they could be conveniently ignored. Unfortunately history has caught up. Fiction writers of today may use the concept of high-energy particle beams as weapons. Unlike their counterparts of the past, these they cannot shrug off. Their writings in this case could be more representative of prophecy than fiction! What is especially disturbing is the fact that such weapons could so easily one day be used against us, not by aliens from outer space, but by some of our fellow men. It is becoming increasingly clear that the dreams most likely to be realized today are those of Armageddon. A recent report by Dr. Lars Erik de Geer, of the Swedish National Defense Research Institute, suggested that the detection of otherwise unexplained neptunium–239 and molybdenum–99 might have originated from the testing of a charged particle beam weapon at Semipalatinsk in southeastern Russia. Both neptunium–239 and molybdenum–99 are *fission* products, however. A beam weapon is *fusion*-based. In the circumstances the report is therefore downgraded. Professor Joseph Rotblat suggested instead that it was more likely that the USSR had succeeded in bringing about fusion by the use of high-energy ion or laser beams. In this case it would be more reasonable to assume a small *fission* fraction to the power source. But, the neptunium and molybdenum could hardly be produced and vented into the atmosphere without the accompaniment of some *fusion* debris.

A very old friend among devotees of science fiction is the almost legendary disintegrating beam. This is not to be confused with the heat ray, laser beam, or high-energy particle beam. Any of these directed against a space vehicle, orbiting satellite, or space station could indeed lead to the disintegration of the object concerned by virtue of explosive decompression. A true disintegrating beam, however, would have the power of causing metal itself to disintegrate—presumably by the rupture of its molecular structure in some way.

It can be argued, and a very good case made, that a heat ray, assuming the temperature it can attain be sufficiently high, is really a form of disintegrating beam since the effect of heat on a solid body is for the average energy of its molecules to increase. As a consequence, the range of molecular excursions from positions of equilibrium also increases. When a certain temperature is reached, the molecules of the material stray so far from their equilibrium positions that they do not return. Instead, they fall into a new position of equilibrium and oscillate about this. As temperatures continue to rise, the energy of motion of the molecules eventually becomes so great that there is *constant rearrangement of position* going on among them. At this point the substance has attained a completely liquid state. The tensile strength inherent in a solid has gone since the molecular structure of the material in the solid state has, in effect, disintegrated.

Under normal conditions there exists an attractive force between individual molecules in a solid. It is this force that keeps them in place and imparts to the solid, of which they are the substance, its strength. It is the gradual loosening or diminution of this attractive force resulting from continued rise in temperature that finally leads to the liquid state. If in some manner a beam illuminating a hostile space vehicle in its entirety for a fraction of a second could negate this attractive force, the effect would doubtless be castastrophic for the space vehicle. Suggesting such a power is, so far as we are concerned, the purest speculation. Whether or not it would be feasible to a technology far ahead of our own, we cannot tell. Certainly such a weapon would be devastating in the extreme.

Nor is it possible at the moment for us to come up with ideas that might one day lead to the magnificent blue, green, yellow, and violet rays of the space opera of science fiction. This may or may not be a source of regret. It all depends on the viewpoint! Such rays are exciting when portrayed on the cover of a sci-fi magazine; the reality would perhaps be more disconcerting. Multicolored rays with the exception of the red laser, may belong to the world of make-believe forever. The really lethal rays would probably be invisible—and perhaps because of that, more insidious and more dangerous. But after all, does it really matter what color is the ray that kills?

14. ASTROBIOLOGICAL WARFARE

Biological warfare is not a pleasant concept. We must, however, consider the possibility of germ or biological warfare being one day used against us by invading, intelligent creatures from outer space. In H. G. Wells's *War of the Worlds* we saw how in the end the unpleasant, octopus-like Martians were defeated, not by the resources or technology of terrestrial civilization, but by the small organisms to which man had become resistant over the ages. Such a sequel provided an interesting and dramatic climax to the book. It also represented a most convenient way out of an impasse for the author, since by then Wells's Martians had gotten a firm hold upon our planet.

This slant by Wells was interesting in another context. The Martian invaders fell victims to the germs of Earth. But the people of Earth might also presumably have succumbed to the germs of Mars inadvertently carried here by the Martians. Either way, this is not germ warfare. At least it is not deliberate germ dissemination. But unless very strict precautions are taken—and even these might prove inadequate—this sort of thing might very easily happen. What was envisaged (or might have been envisaged) when *War of the Worlds* was written could presumably take place during any future contact between our race and another from the stars. A deputation of friendly aliens from, for example, a world of the star Epsilon Indi might, in a matter of hours or days, have been mortally stricken by germs on Earth to which, over millions of years, we had developed a

total resistance. What a tragedy and a travesty this would be—
representatives of a cultured benign civilization capable of reaching us from the environs of a sister sun wiped out by lowly
things! With equal ease we could soon start succumbing by the
millions to a disease inadvertently brought to us by these same
friendly Epsilon Indins. Worse still, both could occur. Swapping
germs with aliens could become a very lethal business.

In no sense do these eventualities represent biological warfare
waged between planets. It is interplanetary infection, quite a
different matter. But the *results* might not be all that different.
Biological warfare in the full and accepted sense of the term is
really an absolute weapon easily capable of destroying entire
populations either directly or by initiation of self-perpetuating
disease epidemics. In the first instance we will have a look at the
concept and its implications. We can then proceed to the extraterrestrial connotations such as they are—or as we believe they
might be.

Any nation having an up-to-date scientific and technological
capability can produce effective pathogenic agents in adequate
quantities. With these available it then has three options open to
it (assuming total disregard of all humanitarian and moral principles).

1. It can attempt to incapacitate a limited number of key personnel so as to delay or inhibit military, civil, and industrial
   preparedness against attack.
2. It can use pathogenic agents offensively as an adjunct to
   conventional attack.
3. It can wage biological warfare so that efficiency, productivity,
   and morale of the opposing side are seriously lowered.

The use of biological weapons can therefore be either open or
surreptitious. A nation adopting the former course is immediately open to payment in kind unless the opposing side is less
scientifically advanced and lacks the necessary means of delivery. A surreptitious use of biological weapons on the other hand
can be part of an undeclared war. The side attacked might not
even be aware that it had been attacked.

We have already spoken of the effect of terrestrial germs on Wells' hypothetical Martians. We could, however, have pointed to classic instances that have already taken place. Bubonic plague (the much feared "Black Death") thinned the ranks of the ancient Crusaders far more than did their enemies. During the South African war of 1899–1902 typhoid fever accounted for the lives of more British soldiers than did the bullets and shells of the opposing Boers. Probably the most noteworthy and appalling of all is the outbreak of so-called "Spanish" influenza that decimated military forces and civilian populations alike during the closing years of World War I (1917–1918).

Earlier we remarked that thus far in terrestrial history germ warfare had not been used. This is true, but only in the sense of a deliberate, scientifically planned and organized attack using disease germs whose lethal propensities are known and clearly understood. Crude attempts at germ warfare *have*, in fact, been made from time to time. These indicate all too clearly that terrestrial man has long perceived its value as a weapon of war. In early times the bodies of cholera and plague victims were sometimes dropped or catapulted over the walls of besieged cities. Infected corpses were also at times dropped down wells supplying drinking water to an enemy, or left strewn about an area that the opposing side was about to occupy. Napoleon Bonaparte is reputed to have deliberately flooded the ground near the beleaguered city of Mantua in the belief that this might accelerate the spread of malaria among his Italian adversaries. If the female of the anopheles mosquito were around in sufficient numbers to utilize the stagnant pools so created, the wily Corsican probably succeeded in his aim!

To be an effective biological weapon a disease agent must possess certain specific features. These are (1) a high degree of infectiveness; (2) considerable powers of resistance to such forces as heat, sunlight, and drying; (3) the capability for rapid dissemination; (4) the capacity to cause high initial mortality among its victims. In addition, the disease agent must be foreign to the area against which it is used so that natural immunity against it has not built up.

Biological weapons embodying known terrestrial diseases

could be regarded as varying in their efficiency. For example, cholera, typhus, smallpox, and similar diseases are appropriate if the deaths of large numbers of the opposing population are desired. The likes of brucelosis on the other hand would probably cause only large-scale sickness greatly weakening powers of resistance to the attacker, of course, and straining the medical resources of those attacked.

Pathogenic agents could, as far as we are concerned, be spread by a variety of methods. These could result in contamination in a number of different ways, i.e., via bodily contact, air, food, or water. The means of dissemination might be by bombs dropped from aircraft, by aerosols sprayed from aircraft or rockets, and even by long-range artillery shells. The last named may seem a little outmoded today, yet had the Germans cared to load disease germs instead of high explosive into the shells fired from "Big Bertha," their incredible monster cannon, during World War I, could Paris have been infected from seventy-five miles distant? The first victims of any pathogenic attack would obviously be those affected directly by the biological agent used. A secondary wave of victims would then ensue by infection from those originally infected, and so on. Once this particular wheel started spinning, there is no saying how and when it would eventually stop.

One of the most difficult problems for the defense in biological warfare is the rapid detection and identification of disease outbreaks, since neither microorganisms nor their toxic products can be detected by any of the senses. Detection and identification can take several days. By that time much can have happened—all of it unpleasant.

Protection is difficult to achieve and total protection simply impossible, except for relatively brief periods. Protective measures involve the use of individual face masks, protective shelters, airtight clothing. Clearly it would be impossible for persons to conduct a normal everyday life under such severely inhibiting conditions.

So far as military applications are concerned, biological weapons fall into the tactical category, since the user might (depending on the germ) have to wait days or even weeks for the full effect to become apparent. But he could afford to wait, for in essence the germs would already be attacking and doing his work for him. When the opposing side was decimated and weakened, the attacker could move in. This move would have to be carefully made, however, since the infections spread by biological weapons do not distinguish between friend and foe.

Such weapons are potentially so insidious that months or even years might elapse before the actual outbreak of hostilities took place. The fighting strength and resolution of a defender would by then be grievously undermined.

This then, briefly, is the general background to germ warfare as seen by us here on Earth. Its use by another galactic society against either ourselves or the inhabitants of some other planet need not, and very probably would not, conform to these basic essentials. Indeed, a number of factors would seem to downgrade its use in favor of some of the more spectacular weapons and techniques discussed in previous chapters.

1. A tactical, slow, insidious germ attack against Earth's population would be totally unnecessary if victory could be secured swiftly and easily by the use of lasers, heat rays and the like.
2. Alien beings proposing to use germ warfare against us would first have to be aware whether the particular pathogenic agent they proposed would prove fatal or merely debilitating. Though apparently virulent to them, it might prove noneffective against us.
3. They would also have to know whether conditions on Earth would or would not inhibit or negate the effect of the particular germ.
4. The use of germs lethal to the aliens and therefore very likely to us also could easily lead to a situation in which the aliens themselves fell victim to the attack they had unleashed—a supreme example of poetic justice!

With this last fact in mind, a potential alien invader desiring only our planet and not us might elect to spread his lethal pathogenic agents while remaining at a safe distance from the stricken planet. Indeed, if they had the power, they might prefer to send the germs in robot probes that burst on entering our atmosphere or in crashing to the ground. They could then wait, perhaps for a year or two, while Earth's population died. When eventually they did descend upon our world, the disease would have died out in the natural course of events. Such an appallingly cold-blooded policy is horrible beyond belief, but as by now we have good reason to know, the end is frequently regarded as justifying the means, even here on our own planet.

Is there any pathogenic agent that, to our knowledge, could really bring about the total cessation of active life on this planet? One that springs to mind at once is bacillus botulanis—one of the most dangerous toxins known. It acts upon the nervous system. Disturbances of vision are followed by muscular weakness and then death due to respiratory failure. It has been calculated that an absurdly small amount of this toxin could wipe out the entire population in a relatively short time. So far as alien attack is concerned two questions at once arise.

1. Would the aliens be aware of the deadly nature of this toxin with respect to the peoples of Earth?
2. Could the aliens produce it?

It seems reasonable to assume that another galactic race wishing to occupy our world would probably be similar to ourselves in a number of ways. On their own world this same deadly toxin, or variants, could be possible. The rest follows. And even though they were not aware of botulanus, they might well have something even more deadly. Advanced technology also implies advanced biological knowledge and expertise.

On balance, however, the launching of biological warfare upon us by an alien civilization appears much less likely than a straightforward open attack using some of the methods and techniques we have considered in the preceding chapters. We could, however, and this seems well worth repeating, be attacked by germs brought inadvertently by the invaders to which they have long been rendered resistant. Conversely the invaders could die because they lacked the necessary degree of resistance to bacteria to which we had long developed immunity.

This aspect, which we might aptly describe as one of interplanetary cross-infection, would, if we ever come into direct physical contact with aliens, have to be considered very seriously indeed. There already exist parallels here on Earth. Not infrequently disease to which civilized people had developed a certain measure of resistance have ravaged and decimated simple tribal peoples when the two came into contact. In no sense of the word was there a deliberate attack but the effects were no less unpleasant.

Though our theme has been one of interstellar war and interplanetary aggression we hope fervently that our contacts with alien beings, when and if they occur, will be to our mutual benefit and advantage, that peace and harmony will reign in our corner of the galaxy. If the galaxy contains a high proportion of advanced civilizations almost inevitably some are going to be of the totalitarian, heavily armed type whose essential doctrine is that might is right. We must hope and pray that our cosmic paths will never cross. But it would be most tragically ironic if a friendly contact between peoples of neighboring stars were to prove catastrophic simply because they passed on to one another germs of a highly lethal nature. Just as we have strict medical control and quarantine regulations on our planet, so also will they be required by other planets in an era of increasing interstellar travel and contacts. We cannot quarantine an entire galaxy.

Recently, the question of infection caused inadvertently by the arrival of meteors or by Earth passing through regions of space where somehow germs had persisted has been occupying the minds of certain biologists. The present writer is not qualified to pass judgment on what seems a highly novel concept. Presumably only time and patient research will produce a definitive answer. Should the theory be proved, the question to be asked next would presumably relate to how such lethal organisms came to be in meteors—or in attenuated "gas" clouds through which Earth may pass. Could either of these things be relics of astrobiological warfare of the past between alien civilizations now long dead? Who can tell?

15. FORCE FIELDS

In the years immediately preceding the outbreak of hostilities in Europe in 1939, there was much talk about the possibility of some kind of ray or beam, the effect of which would be to render the internal combustion engines of contemporary aircraft inoperative. Little evidence of much practical research toward this end can be found, however, and it seems fairly safe to say that the idea never really got off the ground. The technology of the day was simply inadequate, although with war clouds gathering, the need for such an invention was obviously very great. The wish for such a device or technique was really father to the thought. This is hardly surprising in view of the feelings engendered about air attack at the time, especially those concerning the bombing of large, densely populated metropolitan areas. World War I had mercifully come to a close before this sort of indiscriminate slaughter could really get into its stride. True, there had been German Zeppelin raids on London and on one or two other places along the east coast of England, but, comparatively speaking, they had done little damage. The menace at that time was as much psychological as physical, probably even more so. At first the huge Zeppelins had things pretty well their own way. They could come in at altitudes that pursuit planes of the period either could not reach or could reach only slowly and with very considerable difficulty. These great ships of the air could hover silently above the clouds with their motors stopped and no one down below would even be aware of their presence—until bombs began to whistle down, apparently from nowhere. This trick was not always without danger to the Zeppelin crews since at the altitude they flew, low temperatures could congeal the engine oil so that the motors would not restart. And an airship drifting helplessly without its motors is a less than healthy spot in which to be. Indeed something has to be said for the German airship crews. Though popular feeling ran high against them in England at that time (the press of the period had christened them "baby killers") these early German aviators were called upon to endure tremendous physical hardships—long hours of exposure to the most intense cold, the effects of rarified air, and the ever present knowledge that they were supported in the sky by millions of cubic feet of highly inflammable hydrogen gas. One spark, one incendiary bullet through just one of the great gas bags within the ship, and it became a flaming, falling funeral pyre from which there was no possible hope of salvation—only a glorious Wagnerian end! After a while, newer and faster British pursuit planes got the measure of the Zeppelins and were able to pour streams of incendiary bullets into the huge gas cells and their highly inflammable contents. Exit the Zeppelins in a blaze—though hardly one of glory! One British airman even flew a few feet *above* one of these great ships of the sky and calmly dropped a bomb into it. The result was indescribable. It lit up a hundred square miles of the countryside far beneath!

The Zeppelins were succeeded in attacks on London by the first German long-range heavy bombers—the equally notorious Gothas. These promised to give more trouble than their huge predecessors, but before they could achieve this Germany had sued for peace. By then the British air force had also developed a heavy bomber of adequate range to mete out similar treatment to Berlin and other German cities and industrial centers.

The result of all this was that the nations of the world, especially those in Europe, developed a considerable fear of the air weapon used in a ruthless and indiscriminate way. To some extent this fear paralyzed the British and French governments of the time and prevented their taking positive action against the rise and increasing aggressiveness of the European dictators.

Films, books, and radio commentaries made much of future aerial warfare, extrapolating aerial bombardment from its sudden cessation in 1918 and almost certainly endowing it with greater powers of destruction and horror than it then possessed. A notable example of this and one which very much epitomized the feelings, fears and thoughts of the period was the film version (1936 or thereabouts) of H. G. Wells' *Shape of Things to Come.* The opening scenes of the movie were filled with horrific shots of air raids devastating a great city.

This kind of thing might have been passed off had certain all too real events of a like nature not then been taking place. In 1931, Japan had wantonly attacked China and given a fair example of what the bombing airplane could do to undefended cities. In 1935, Mussolini's forces invaded Ethopia and further proved the point, adding poison gas bombs to the more conventional high explosive and incendiary variety. Came 1936, and the power of the aerial weapon really began to be seen in the attacks made against open (i.e., undefended) cities during the Spanish Civil War. To this day the name of Guernica is synonymous with carnage and destruction wrought from the air. And by 1937 the Japanese were further improving the technique in China. Until that time Hitler's vaunted and much feared Luftwaffe had not shown its powers (other than semiclandestinely in Spain) and Europe quite literally shivered at the thought. When Britain and France finally declared war on Germany in September 1939, because of Hitler's invasion of Poland, most people feared the immediate advent of terrible and devastating air raids. The writer, then a schoolboy of sixteen, remembers as though it were only yesterday, the night of September 1, 1939, when a total blackout first became compulsory and the probing silvery beams of searchlights began to sweep the dark, star-strewn skies. At that age excitement far outweighed fear. It was as if the strange world of science fiction and the future had at last arrived. Was this not H. G. Wells' *War in the Air* and *Shape of Things to Come* about to materialize? This, then, was the atmosphere in which the concept of a motor-stopping beam brought a strange comfort. The equivalent today would probably be a beam that could render nuclear bombs inoperative.

How wonderful it seemed back in the late 1930s if only such a device could be perfected. At once all the raiding air fleets, real and potential, would be rendered useless.

Press a few buttons here and there and the great black bombers with their loads of death and destruction would be forced to crash land, their motors dead. From time to time reports in the press of the period spoke of experiments along these lines. Even Marconi, prior to his death in July 1937, was said to have been working on the idea, although little detail of his research has ever come to hand.

Basically, the concept appears to have centered on some sort of electromagnetic beam that could inhibit the magnetos and other essential electrical components of aircraft motors. It must be stressed that had such an invention proved possible at the time, its efficiency might have been much more limited than was popularly supposed. Whereas it would have proved highly effective against aircraft using petrol engines, the same would not have been true with respect to machines with diesel engines. The latter do not require an electrical ignition system and would have gone on functioning just the same. Unhappily, most of the Henkels, Dorniers, and Junkers of Hitler's air fleets used diesel engines.

So far we have only been considering the past as background. It is time now to move forward and try to place all this in the context of a future interplanetary or interstellar war.

We have seen the attacking aliens as belonging to a civilization so far ahead of our own in technological knowledge and expertise that the defending weapons we could use—aircraft, tanks, chemically-propelled spacecraft—would be swiftly and ruthlessly swept aside by ultra-sophisticated weapons of the most lethal kind, such as disintegrating beams. Let us suppose, however, that though the invaders are undoubtedly our superiors (clearly they must be to have brought a task force to the Solar System from another star), their degree of superiority over us is somewhat inhibited by the fact that in *quantity* of equipment we might have the advantage. In such circumstances our aircraft and our armored fighting vehicles *could* represent a certain danger to the aliens. For the record, there are several instances in terrestrial history in which small, well-armed bands of troops have been wiped out by hordes of savages wielding primitive weapons. However, these "primitive" weapons of ours would be no menace to the invaders if the invaders had beams rendering petrol, oil, and jet motors inoperative. The day this happens the teeth of any contemporary terrestrial force are well and truly drawn. An air force is useless and an army is reduced to masses of infantry (or cavalry) at best. Transport of guns, munitions and supplies again depend on the horse, the mule, or the steam engine. If rocket motors could somehow be rendered equally ineffective, then the delivery of nuclear weapons by long, medium, and short range ballistic missiles would be rendered impossible. Only by the use of artillery could an atomic punch be delivered. Unfortunately, artillery pieces capable of discharging nuclear shells are large, heavy, and cumbersome. They could not be moved swiftly no matter how many teams of horses were used. Something in the nature of steam tractors would be required. Such weapons would probably have to be mounted on large steam-hauled railway cars. Their mobility would thereby be greatly reduced, however, and in any case railroads are easily cut and otherwise put out of action.

Even the vanguard of a force very much our superiors could initially find itself at something of a disadvantage if attacked resolutely by vast swarms of the most modern jet aircraft the moment it reached the surface of our planet. This, the bridgehead period, is always one of acute danger to an invading force. It is true that in the film version of *War of the Worlds* absolutely nothing that tanks and aircraft could do against a small bunch of Martians was of any avail. But that after all was a movie. Conditions would not necessarily be the same.

We must admit that so far as we are concerned, there is nothing immediately in sight that could render inoperative the propulsion systems of tanks, jet aircraft, ballistic missiles, or space vehicles. In more ways than one this may be a source of regret, for the existence of such a capability would undoubtedly make our world at the present time a saner and a safer place.

Could this power have been developed by civilizations far ahead of us technologically? This possibility, like many others along these same lines, cannot be ruled out. Certainly an electromagnetic beam capable of fouling up the electrical systems of gasoline-driven aircraft and land vehicles seems a very distinct possibility. In the case of gas turbines and pure rockets the prospects, so far as we are concerned at the present time, seem considerably less promising. This is about all we can safely say on the subject meantime.

At this point we should devote a little time and thought to a related aspect, one that has proved of great use to science fiction writers for several decades—the legendary, or dare we say notorious, "force field." The force field to the fiction writer is one of those delightfully vague schemes that allow the most improbable things to happen with ease. There must be few of us by now who are unfamiliar with the general outlines of the idea. It was illustrated quite dramatically in the film version of *War of the Worlds*. When all weapons had proved futile against the Martians it was decided that a thermonuclear bomb of great power would be dropped by a U.S. Air Force B-52 on a strong concentration of Martians reposing in the hills of southern California not far from Los Angeles. Flying at a great altitude, the B-52 looses its thermonuclear "egg" on the Martians. The resultant explosion (thanks to the effects department of the movie studios) is breathtakingly and awesomely violent. Onlookers many miles from the scene are blinded by the flash and hurled to the ground by the immense shock waves generated. The Martians presumably must have been vaporized in an instant. Alas, not so! When the mushroom cloud has climbed into the sky and the debris and dust have settled, the Martian encampment appears through the smoke and swirling vapor, totally undisturbed and unharmed. Clearly in view is a great dome covering the entire area that they occupy. Normally this, the perimeter of the force field, would be invisible, but on this occasion it is rendered visible by the smoke, dust, and vapor, which obviously cannot penetrate it. Without a doubt such a power on Earth today would make all of us sleep much easier in our beds at night. It is a facet of man's "progress" on this planet that thermonuclear tipped rockets (or MIRVed clusters of projectiles) are guidance-programmed on most of the world's major cities and metropolitan areas. How wonderful it would be were we able, at the flick of a switch, to throw an invisible, impenetrable umbrella over these areas.

No doubt readers will recall many less dramatic uses of the now almost legendary force field. A firm favorite is the invisible shield that arises to protect each and every alien from the evil machinations of terrestrials into which the aforesaid terrestrials blunder painfully each time they approach grounded aliens or their ships. After a time they get the message.

For any writer of science fiction it is easy to talk airily of force fields and leave it at that, but if we are going to credit advanced extraterrestrial civilizations with the ability to produce force fields having a potential along the lines we have just described, it is reasonable to suggest how they might have achieved this, or at least along what lines research to this end might be directed. At once we come up against a most formidable snag, for the truth is that, at the present time, we simply do not have a clue. We may have watched the starship *Enterprise* in "Star Trek" save itself from destruction a score of times by switching on its "protective shields" and have grown to accept this procedure as standard on the sounding of a "red alert." We are not asked to visualize the technique, which is just as well.

It is essential to first look at the concept of force. Modern physicists are agreed that natural forces fall into four apparently distinct categories. We are all reasonably familiar with two of these, gravitational and electromagnetic, since they form part of the background of our everyday lives. All the forces encountered during ordinary experience belong to one or the other of these categories. Even the forces holding an atom together, the forces binding atoms into molecules and molecules into liquids or solids are electromagnetic. When science fiction writers pull that other old favorite out of the hat, the disintegrating ray, they are in fact suggesting a beam of radiation capable of negating these forces so that the molecules comprising the matter of solid-material objects simply fly apart. Here on Earth we may not yet have achieved this power, but we can at least see how theoretically it is possible. The practical problems remain infinite. Not so perhaps to alien civilizations from the stars.

The other two types of force are believed to exist only on the subatomic scale. The forces that bind atomic nuclei belong to the so-called *interaction* class. These have a very short range but are extremely powerful—as is clearly indicated by the fact that nuclear binding energies are millions of times greater than either atomic or molecular binding energies. Thus a disintegrating beam designed to rupture a *nucleus* is a different proposition from one aimed at breaking up matter by destroying the forces holding atoms or molecules together. The remaining category is represented by the weak interaction force responsible for the radioactive beta decay of nuclei.

These four categories of force can be ranked as follows in order of *decreasing* strength: nuclear (strong), electromagnetic, nuclear (weak), and gravitational. Gravitational force is by far the weakest, although it was the first to be recognized as a force. Its gross effects are a function of the enormous masses of the heavenly bodies; for example, the planets move around the Sun because of the very high gravitational power of the latter. Similarly, an object dropped here on the surface of the Earth falls to the ground, not because there is nothing to support it, but because of the gravitational attraction exercised by the earth.

Unfortunately, these definitions and brief descriptions of the four force categories do not in themselves take us very far in trying to conjure up the type of force fields so freely employed in the pages of science fiction. We would assume, however, that nuclear forces would not be involved (and how wrong we could so easily be!). This leaves us with the electromagnetic and gravitational variety.

We can probably best approach the problem—and it is admittedly a considerable one—by considering the common, everyday phenomenon of magnetism. Let us consider two quite ordinary permanent bar magnets. One is fixed to a base with its north pole *uppermost*. Suspended above this magnet, but free to move, is an identical bar magnet with its north pole facing *downwards*. The latter is lowered in an attempt to make the two meet end to end. It is soon apparent, however, that this simply cannot be done. As the upper magnet descends, it is increasingly deflected away from the lower one. The two respective north poles cannot be brought into contact because of a distinct repulsion effect. There is, in fact, a magnetic force field between the two that prevents this from happening. This field is permanent; it cannot be switched on and off at will.

If, on the other hand, we take two small electromagnets and duplicate the arrangement, we find that so long as the current is switched *off,* the two magnets can be brought end to end without difficulty. Switch on the current, however, and at once the upper (freely suspended) electromagnet is deflected away from the lower. The magnetic force field in this instance can be switched on and off at will.

These simple experiments illustrate the fact that electricity and magnetism are in some way closely interrelated. It also confirms a basic physical precept: Among the various material particles in the universe some are positively and some are negatively charged, those of similar polarity repelling, as we have just seen, and those that are dissimilar attracting.

Stationary electric particles are said to be surrounded by an electric force field and moving particles by electric *and* magnetic force fields. This is just another way of saying that if an electric particle is placed at any point in space surrounding one or more electric particles, it will be acted upon by a force by reason of its electrification. The part of the total force depending only on the charge of the particle and not on its velocity is termed the *electric* force. The remaining part, depending on the velocity of the particle as well as on its charge, is termed the *magnetic* force.

The fundamental truth emerging from all this is that an electromagnetic force field can be produced and, depending on the polarity of the two, will either repel or attract. Returning to our electromagnets, we can say that the strength of the field is directly proportional to the strength of the current flowing and inversely proportional to the distance separating the magnets.

This particular force field applies only to metal—and ferrous metals, at that. Moreover, both magnets must have their like poles facing one another, i.e., north to north *or* south to south. If a north pole approaches a south pole or vice versa, the very opposite effect takes place, i.e., there is attraction instead of repulsion. In this we might have something for the basis of the *tractor beams* of science fiction whereby one spaceship is capable of pulling another towards it. The general picture of a force field is now becoming evident.

The fabulous force fields of science fiction, however, do not just affect ferrous metallic objects of specific polarities. In their effect they are almost magically universal, inasmuch as all types of material are repelled—metallic and nonmetallic, organic and inorganic. In most science fiction the effect of a force field is generally portrayed as an invisible *wall* rather than as a *region* of repulsion. This is obviously more spectacular but seems much more improbable. A force field, we would expect, would indicate its presence only slightly at a distance, but as an approach is made towards its source, the repulsive effect would become steadily stronger until a point is reached when no further progress could be made against it.

A magnet will not repel a block of wood or attract it either, for that matter, but if we allow the block of wood to go unsupported, it falls to the ground. In this instance the Earth is acting as the "magnet" and by virtue of *gravitational* force has attracted the block of wood. Unfortunately we cannot apply the converse here, i.e., we cannot induce the Earth to repel the block of wood instead. Neither can we devise an apparatus to do this. The day we can, we have indeed produced our protective repellent force field!

Gravitational fields rather than electromagnetic fields would nevertheless seem to represent a better basis for a repulsive force field. We saw in an earlier chapter how gravity is, in fact, produced as a result of mass. The greater the mass of a planet, the greater is its gravity and vice versa. Thus on the Moon we would only weigh about one-sixth of what we do here on Earth, i.e., the Moon would have a pull on our bodies only one-sixth of that which Earth exerts. On Jupiter, a much more massive planet, the converse would be true, and we would probably be crushed by the weight of our own bodies. If we go to the far extreme and consider a collapsing star, we find a singularity in a black hole of infinite gravitational attraction. Even the photons of light cannot escape from this. Note, however, that in every instance, the force is invariably one of *attraction* and never of repulsion. Gravity seems almost to have something in common with time—a capacity to proceed or act in one direction only!

From here on, like it or not, we are again forced into speculation as to what hypothetical civilizations many millennia ahead of our own might have achieved. It may well be that as yet we do not really comprehend fully the fundamentals of gravitation, whereas some of them have gained a much clearer insight into it. We are capable of creating magnetic force fields with powers of attraction *or* repulsion. Perhaps our cosmic superiors can do the selfsame thing with respect to gravitational fields. If they can and at the same time are capable of giving these fields sufficient strength, then undoubtedly they have their repelling force fields and can presumably protect their spaceships, other craft, their persons, even perhaps their entire home planets, most effectively. What is even more likely is that they can achieve this by virtue of the application of what we presently term "unified field theory."

Unified field theory is rather a complex physical matter and one difficult to compress into a few lines in a book of this sort. It represents an attempt to extend the general theory of relativity (in itself a rather complex affair) to electromagnetic forces and the forces existing between nuclear particles. General relativity incorporates the *gravitational* field into the structure of four dimensional space-time (the four dimensions being length, breadth, depth, and time). Nothing can exist without the essential fourth dimension of time to exist in. The unified field theory attempts to extend this treatment to the other forces mentioned. By so doing, all the fundamental fields of force could theoretically be described by the geometry of space-time. Grossly oversimplifying this adds up to the possibility or even considerable likelihood of all the forces we have mentioned being interrelated in one way or another. It is also possible that the unified field theory might explain some of the difficulties inherent in classical magnetism, e.g., why elements of like charge should repel one another and vice versa. The late Albert Einstein himself believed that the whole of physical reality could be represented by means of a field in which matter becomes simply a region of high field intensity.

We begin therefore to see faintly into a realm of physics of a very fundamental kind and the possibility of revealing, eventually, far-reaching truths, the interpretation of which could greatly alter our technology and indeed our entire civilization. Consequently it is reasonable to suppose that there could be those in the galaxy who have already trodden this path and are therefore in possession now of powers that as yet we see no means of acquiring. Certainly, force fields represent a distinct possibility and might, in fact, not be so very different from those freely portrayed in the pages of science fiction. The protective force shield is probably already at hand to protect alien starships at the touch of a button. Application of the same technology could provide the basis for the power-unit inhibitors dealt with earlier in this chapter. Regrettably, at the present stage of our fundamental physics and of our technology it is impossible to be more precise. If it were, then presumably we ourselves would already be well on the way to acquiring these seemingly miraculous powers. For us this lies in the future. To those aliens with these gifts, their potential in space war is immeasurably greater.

It is important, too, that we do not see exploitation of these theories merely as a means by aliens of attacking and subjugating Earth. If the course of history allows us in time to send interstellar ships or fleets into deep space among the stars, we could easily find ourselves embroiled and attacked by alien ships having these powers. These are battles we could not win. And the first Earth colonists on suitable planets of other star systems could find themselves faced by similar alien attack. Presumably aliens would be no more kindly disposed to our intrusion into their cosmic space than we would be to their entry into ours.

16. TRICKS, TACTICS, AND STRATAGEMS

We have so far restricted ourselves very largely to an actual onslaught on our own planet. Since our dominion at the moment extends only over Earth, this is hardly unreasonable. Despite six highly successful expeditions to the Moon by American astronauts, we cannot as yet truly say that we exercise total jurisdiction over our satellite. When in a few decades permanent stations on the Moon are developed and a terrestrial presence maintained there, we will at last be in a position to say this is so. Within the next century terrestrial dominion should have reached well out into the Solar System—assuming that we haven't ended all existence here on Earth by engaging in nuclear war. We can confidently expect permanent stations on Mars, perhaps even settlers living under the plastic domes and in the environmentalized conditions so beloved by generations of science fiction writers. Temporary settlements even on some of the moons of the giant planets Jupiter and Saturn are a distinct possibility. And, of course, space stations in orbit around Earth and perhaps around some of the other planets are virtually certain, though the purposes of these may be as much military as scientific. In short, by 2079 A.D. man will have come to regard himself more as an inhabitant of the solar system than merely of Earth. Removal of his earlier parochialism will be no bad thing.

Suppose it is at that stage in our history that the presence and threatened encroachment by an alien culture from beyond the

Solar System becomes apparent. Since by then we would be well spread out, the potential battle zone would be more or less the entire Solar System and not merely the immediate environs and surface of Earth. Indeed, it is in this context that we can begin to realize the true dimensions of space war.

Let us extrapolate into time somewhat further by supposing that two or three centuries more have elapsed. By then, space-ships from Earth and the Solar System would be capable of ranging beyond the latter and reaching some of the nearer star systems such as Alpha Centauri, Epsilon Indi, Tau Ceti, and Epsilon Eridani. Such a time scale should not be seen as a prediction. Interstellar travel by terrestrials could be achieved by then; on the other hand, it may take much longer.

When ships from Earth and the Solar System start nosing around the planets of other stars, they could easily stir up a hornet's nest. On such occasions *we* would be the intruders. Alternatively, we could bump into other intruders from else-where with the same idea in mind. That hornet's nest could become equally lively! Again the contact in space could come centuries earlier when, as already mentioned, we found alien ships trespassing over the orbit of Pluto.

What follows from any of these eventualities, if interstellar war results, is the almost classic science fiction scene—battles between fleets of space ships, battles on and around space sta-tions, raids and running fights on some of the satellites and asteroids. Indeed this all seems so sci-fi in concept that a writer is tempted to cry "halt!" before the subject runs off completely into that medium. Yet this would be wrong. A little thought brings the realization that science fiction writers have to a very large extent only been visualizing the future. There is no real reason to suppose that the form and sequence of events should be so very different from those predicted. When we think about it, the first lunar landings ran fairly true to prediction, the outstanding difference being the fact that the whole moonship did not descend to the lunar surface. We cannot, of course, endorse all the peculiar effects and phenomena dreamed up for our entertainment by the producers of certain space movies. Multicolored rays that cause opposing ships and men to glow brightly, then disappear, still appear highly improbable. A heat ray that turns living creatures into sticky, carbonized blobs on the landscape may be less aesthetic but much more likely.

Before we get around to thinking of space dreadnoughts firing broadsides at one another with lasers, heat rays, disintegrating beams, or what have you, let us dwell a little on the theme of raiding parties and small forces on some of the satellites, of representatives from opposing sides meeting one another on the lonelier outposts of star systems. We deliberately say star systems because the events we are considering could one day take place between ourselves and alien beings either in our own Solar System *or in others* depending on the time scales we mentioned earlier.

Since satellites and asteroids are likely either to be airless or to contain unbreathable atmospheres, the use of spacesuits (a generic term that covers everything from life-support systems in vacuum to the same in poisonous or unbreathable atmospheres) will generally be mandatory. If such a suit worn by a person in the vacuum conditions obtaining on the Moon or a similar body were seriously punctured, the demise of that person would in all probability be swift and most unpleasant. The blood would boil and the body would very likely burst. (The moviemakers don't appear to have gotten their special effects people on to this one yet, for which we should be thankful.) It is as horrible as that. No doubt self-sealing suits like self-sealing gas tanks in aircraft, could and would be developed, but obviously there must always be a limit to this sort of thing. No gas tank yet, self-sealing or not, is capable of surviving the impact of a projectile that can make a really large rent in it. The same would certainly be true in the case of spacesuits.

The "Achilles heel" in these suits, worn in nonvacuum but nonbreathable atmospheres, would be such parts as the oxygen pack or the delivery tubes. If these, for one reason or another, were to be seriously ruptured, beings used to a respiratory diet of oxygen/nitrogen are not suddenly going to delight in an atmosphere of hydrogen or ammonia. In fact, the event is not going to do the wearer any good at all and he or she will almost certainly depart this life in a most painful manner.

Under vacuum conditions personnel engaging in military operations are going to be under very considerable risk. To stop an enemy—and that very permanently—one has only to puncture his suit sufficiently. That is all. Nature will do the rest. Presumably weapons designed for this role need be nothing more remarkable than ordinary firearms. A conventional bullet or two piercing a space suit, even if they did no harm to the occupant, would nevertheless be responsible for his rapid, decompressive demise. A small gas gun firing large-caliber darts might be equally effective in this role. Especially noteworthy in this respect would be the weapon envisaged in a science fiction story of 1939 vintage. This particular weapon was employed by the occupants of a remote, rather backward planet against a party from Earth. It fired small discs of metal whose rims were razor sharp. It had a very rapid rate of fire and the diabolical little discs whirled through the air in a deadly hail. In the story, the atmosphere of the planet was breathable to terrestrials. Thus no space suits or life-support systems were required. But the discs, with their deadly, high-speed cutting edges made mincemeat (almost literally) out of several of the unfortunate party. The potential of such a weapon used against people wearing space suits in very low pressure or vacuum conditions is grimly obvious. It is equally obvious that such a weapon could be easily developed. Its great potential against space suits would be the fact that the discs could slice large rents in them with catastrophic results for the wearers, unless the exterior of the suits were reinforced in some way. Perhaps the wearers will be armor clad—like knights of old.

Personnel working on the exteriors of space stations or on the outside of space vehicles would be equally vulnerable to projectiles of this or some other kind. Presumably attempts would be made to render the material of space suits less vulnerable to attack of this sort, but clearly there must be limits to what can reasonably be attained. Before, and in the very early days of space activities, the danger from small meteorites was strongly emphasized. Certainly, an astronaut either in space or on the surface of an airless body like the Moon would be in dire straits were his suit to be punctured by the impact of a small meteorite. So far (perhaps a little surprisingly) this danger does not seem to have materialized, though it must still exist. It is now known that the surface of the Moon is being continually churned up by the impact of micrometeorites. Contemporary space suits appear to be resistant to them. It is the occasional larger one that represents the real danger.

The idea of spacesuited belligerents pursuing and attacking one another in the eerie twilight of distant planets and asteroids conjures up memories of those early days when Flash Gordon and others did that very thing. There should probably be a clarification here, since Flash and his friends were always, it seemed, fortunate enough to find themselves on worlds and moons where the atmosphere was breathable and the gravity quite normal. In the circumstances space suits were unnecessary. Perhaps the producers of these epics felt it would be a shame to encompass the rugged Flash and the lovely Dale in such unsightly and unglamorous apparel. But many are the sci-fi tales of this period that featured events along these lines. In considering these things one begins to realize that the future envisaged in these stories may be closer than we then believed. In 1939, such ideas were the purest fiction. Many felt this was all they ever could be. Others, more visionary in outlook, saw them as probable events of the late twenty-first or twenty-second centuries at least, and perhaps even beyond. Certainly no one then would have believed that within three decades the Moon would be reached, that in thirty-eight years we could gape at the barren surface of Mars from the comfort of our armchairs. The tempo and acceleration of space technology and expertise continues to increase, in spite of the works of politicians and economists. And so today, some forty years later, we can think of adventures amid the moons and asteroids of the Solar System and the planets of other star systems not as impossibilities or incredibly remote happenings but as events that could and probably will come to pass within a foreseeable future. It is as though the time scale of events were being squeezed and the distant future sent hurtling toward us.

We are, as usual, thinking of ourselves being involved in these twilight operations on worlds and moons of this and other solar systems. Indeed, this has been the basis of our theme all along. It is right to say however, that the whole concept of space weapons and of space war could be—and probably is already—taking place far from our section of the galaxy. This could so easily have been the position for centuries or millennia. And—most important of all—it could remain the pattern for the future. Clearly, so far as we are concerned, it is an infinitely more pleasing thought. If we are to have contacts with aliens, we want these contacts to be pleasant and productive. If not, we do not want them at all and would far rather continue to pursue the independent if lonely existence of homo sapiens while other galactic civilizations battle for supremacy.

Many of our science fiction heroes and heroines have become involved in star wars far from the realms of their native home—generally in the interests of the good guys and to the eventual detriment of the bad ones. This to some extent parallels the action of those who volunteer to serve in foreign wars on Earth for reasons of principle, adventure, or plain hard cash. Perhaps someday, with interstellar flight well established, these things might come to pass. In this way those who enjoy a star war now and again could be assured of their fun. There would, of course, always be inherent risks in such actions. The presence of members of our race on one side or other in a star war would point to (1) our existence and (2) our interstellar capability. The end product could be punitive raids against Earth for interfering—or a full-scale assault to ensure that such interference did not occur again.

Right now, the prospect of men and women being able to penetrate the stellar depths at will for such purposes remains very much in the realms of science fiction. To us at the present time the concept of interstellar travel is so tremendous (and not a little frightening) that we see it as something that could only be brought about by the combined efforts of the nations of Earth. At first this would undoubtedly be so. At the moment, although it is technically possible for any of us to visit the Moon, no one can go there at will. To reach our satellite is still a tremendous undertaking requiring the resources, skills, and organizations of one of the greatest nations on this planet—the United States of America. Any of us who personally fancy a week on the Moon to relieve the monotony of life down here are unlikely to get farther than the main gate at Cape Canaveral. Yet a day will surely come when trips to Earth's only satellite will be on a commercial footing and available to those with good reasons for going and the necessary cash for the fare. Admittedly this presupposes the existence by then of lunar settlements and colonies in closed environments. If we look still further ahead in time (*very much* further it must be conceded) there could come an era in which planned explorations and expeditions to other solar systems became feasible propositions. Indeed, assuming we have been left untroubled, it could be the peoples of Earth making their first move toward "creeping colonization." Carried to its ultimate and logical conclusion, the slant of all the preceding chapters could be reversed—a case of "for 'us' read 'them' and for 'them' read 'us'!" Terrestrial aggression would then be loose in the galaxy!

So far in this chapter we have dwelt on the prospect of small bodies of men and women on the outposts of solar systems. But space war in its fullest sense is generally envisaged as great space fleets battling it out or as running fights between rival space cruisers. Have them slicing bits off each other with rays of primary colors and we are truly in the world of science fiction—or are we? Might we not just be in the world of the distant future? Delete the "technicolor" rays and substitute something less garish and spectacular and we could well be on the right lines.

What are the tactics of opposing fleets of space cruisers going to be? One is tempted here to draw some sort of parallel with terrestrial naval fleets, though it can hardly be a very valid one. True naval war (at least along World War I lines) was basically a slugging match. The tactics were to get one's ships, by skill or good luck, into a position to do the slugging while preventing the opposition from doing likewise. This form of warfare was a two-dimensional struggle on the surface of the ocean, involving slow lumbering masses of steel hurling high explosive projectiles up to twenty miles. Spacecraft move much more quickly and are considerably more maneuverable; their future weapons will have a much greater range and be infinitely more deadly. The comparison then can hardly be regarded as very viable. A comparison with opposing air fleets next suggests itself. Aerial battles are much more fast moving than their naval counterparts and take place in three dimensions just as space battles would. But the speed of aircraft cannot approach that of space vehicles. One common thread does run through all three, though, for the aim in each case is to gain supremacy over the medium involved so that subsequent operations can be carried out as free as possible from interference or harassment by the enemy; i.e., for "air cover" read "space cover."

There probably is no true parallel. Air war did not mirror naval war, and we can safely assume that space war will parallel neither. Space is a new medium and the tactics of space combat are going to be different—just how different we cannot really say until the thing begins to happen, or until returning star voyagers one day tell us what goes in remote parts of our galaxy. It has often been said that in war, terrestrial war that is, there is a strong and dangerous tendency, at least by some nations, to employ tactics and techniques perfected in an earlier contest. Britain and France did just that in 1940. Their opponents, the Germans, made no such mistake. The rest is history. This is only one instance. There are many others. It would be a mistake, therefore, to see (and far worse to model) space war tactics along the lines of naval and aerial engagements fought many decades earlier. The tactics of space combat will be worked out in the hard school in which all combat tactics are developed—in practice; by trial and error, by ability or failure to learn.

This is not to suggest there can be no parallels whatever. A favorite tactic in naval war was to use capital ships (i.e., battleships and battle cruisers) as a kind of pawn. They need not be used, might in fact never do battle. The mere fact of their presence could be very inhibiting to an enemy. Hitler tied up a large part of the British navy in World War II merely by putting the battleship *Tirpitz* in a sheltered and guarded Norwegian fjord. A similar scheme was tried out by the British just after Pearl Harbor in 1941. The almost total destruction of the United States Pacific Fleet (or at least of its battleships) meant that the Japanese, with an immensely powerful navy, had things more or less their own way since the British navy at that time had its hands full in the Atlantic and in the Mediterranean. It was against this somber and dangerous background that Winston Churchill decided to commit the latest British battleship *Prince of Wales* and the older battle cruiser *Repulse* to Pacific waters so that they could, as he put it, "exercise that kind of vague menace which capital ships of the highest quality whose whereabouts are unknown, can impose upon all hostile naval calculations." His idea was that they should go to sea (from Singapore) and vanish among the innumerable islands. Basically the idea was sound. Tragically, it was not properly tried. Instead the two great warships put out to the open sea with four destroyers and no air cover. For the second time in three days the Japanese disposed of a fleet of capital ships. On that day the era of the dreadnought can be said to have ended. For the United States and for Britain it was a hard lesson.

All this may seem a very far cry indeed, both in space and in time, from a possible interstar war of the future. Let us consider a parallel, however. The British battleships were to lose themselves among the innumerable islands. But where in the Solar System are these "innumerable islands" for space dreadnoughts to lie lurking? In a belt between the orbits of the planets Mars and Jupiter. Here lie the asteroids—thousands of little planetary "islands" in a region of the Solar System so far from the Sun that the light of the latter has become dim. Powerful and heavily armed space cruisers could lurk here in the dim, eerie twilight. By so doing they could represent a powerful threat to invaders from beyond the Solar System, assuming of course that in design and armament they were a reasonable match for the enemy vessels. The incoming aliens prepared to do battle with the terrestrial fleet in the environs of Earth could find themselves unexpectedly attacked from the asteroid belt far from their destination. And also just as great ships can hide themselves at sea if the ocean wastes are vast enough (the Japanese task force that attacked Pearl Harbor was never detected in the waters of the North Pacific), so could space men-of-war disappear into the immense dark wastes between the orbits of Uranus, Neptune, and Pluto.

Readers may be quick to point out that this analogy is less than perfect, since radar and other sophisticated electronic sensing devices aboard the alien ships would surely be able to detect the presence and whereabouts of lurking terrestrial vessels. Space, however, even within the confines of the Solar System, is so incredibly vast that accurate detection along these lines could be difficult. By that time also we would probably have the means to confuse hostile radar. It would be possible as well for defending terrestrial and solar forces to plant "ghost ships" in various orbits. These could be older, obsolescent craft from which all armaments had been removed and which, void of crews, would be remotely controlled. The problem for the invaders would be in determining which was the innocuous dummy and which the real thing.

Let us suppose also that part of the alien invasion force is detached to deal with a terrestrial vessel that has just been detected. We would assume that our fellow terrestrials of the future would be aware of the menace drawing closer to them. Here could be scope for leading the enemy into one of several kinds of trap. Three possibilities suggest themselves. If the terrestrial ship lay near or just within the asteroid belt, it could then take refuge deeper in the zone. Since the asteroid belt comprises a host of bodies ranging from giant boulders to small planetoids, it represents a highly dangerous region in which to engage in pursuit or hide and seek, most of all to aliens to whom it could be a relatively unknown quantity. Thus, alien ships plunging in pursuit of a lone terrestrial craft could be shattered by collision with small asteroids. The second possibility is that of using a lone terrestrial ship as a form of bait. Behind a large asteroid, moon, or planet, a whole squadron of terrestrial defenders could be lying in wait. In such a position they would be effectively screened from the radar sensors of the aliens. Once the "bait" ship had led the invaders to that planet or moon, the defending ships could strike. The third possibility is that the alien ships could be lured into a space minefield strewn suddenly before them by the retreating defenders, an idea we mentioned earlier in chapter 5.

Whether or not the idea of the "Q" ship might be projected into the concept of space war makes for an interesting point. It represents the kind of subterfuge that succeeds initially but rarely thereafter. The original idea of a "Q" ship was developed by the Allies during World War I. At that time German U-boats were proving a very considerable menace. They preferred to expend their valuable and highly lethal torpedoes against warships, escorted merchant ships, and very large merchant craft going it alone (e.g., the ill-fated Cunard liner *Lusitania*). Sinking an enemy merchant ship by gunfire after surfacing was less quick but more economical—as long as no naval craft were around. The "Q" ship looked to all intents and purposes like a lonely, plodding, rusty, old freighter. Up popped the German U-boat to sink it by gunfire, but before the U-boat crew could man their deck guns, covers and hatches on the decks and in the sides of the "Q" ship fell aside to reveal batteries of quick-firing naval guns manned by naval personnel. Their first salvos in many cases sent the U-boat straight to the bottom. The remedy so far as the Germans were concerned was simple: Sink at sight with torpedoes, particularly harmless looking old freighters going it alone.

The parallel in space might apply in a long-running war of attrition between rival star systems. A lumbering space freighter looking as though it might contain valuable supplies is halted by a hostile space cruiser. For its pains the latter is seared suddenly and catastrophically by a battery of concealed ray projectors on the freighter. The remedy would again be precisely that on which the Germans fell back—destroy on sight.

This is a facet of operations we can extend a little further although it depends on one of the adversaries having the tech-nology to perfect matter transmission of living beings (see chap-ter 10). In the context of a war between ourselves and aliens a terrestrial freighter is halted by an alien space cruiser. As the latter draws within matter-transmission range, a force of highly trained terrestrial shock troops is beamed aboard the other ves-sel. With luck, and drawing on the essential element of surprise, the alien vessel could be captured intact. Such a prize could then sail into the middle of an unsuspecting alien force and create terrible destruction at close range. So far as we are con-cerned a trick like that must remain a very speculative proposi-tion, since even the genesis of matter transmission is not yet in sight. All too easily, and more probably, this is a technique that could be used by aliens against our own ships. In such a case our spacecraft would have to keep well out of matter-transmission range of all hostile vessels.

A rather different slant can be given to the idea by transmit-ting aboard a vessel of the opposing side, not a body of armed men, but a primed explosive charge, chemical or nuclear, which could be detonated as soon as the "donor" ship had retreated out of harm's way. A touch on a button, a pulse of RF (radio frequency energy), and————! This could represent a very sinister, insidious, and devastating form of attack, especially if a nuclear device of small dimensions could be implanted in a part of the ship where it was either unnoticed or lodged in a highly inaccessible spot. And even if the device were found and extracted, it might not be possible to defuse or jettison the thing in time. Once again, because of their advanced technology, alien attackers are more likely to possess this ability than we. The possibilities of attack along these lines, if not limitless, are at least considerable.

As we peer myopically into the future through the mists of time, the difficulties inherent in trying to assess the tactics of war in space become steadily more compounded. One of the main stumbling blocks is the degree of technological superiority that one side might have over another. If our system and planet should be invaded by an alien culture in the immediately foreseeable future, the question of our tactics against the aggressor barely arises. Invaders able to reach us have succeeded in overcoming the very awesome parameters of time and distance. We have only reached the Moon. There is really no contest in such circumstances. All we could attempt at best are some of the possibilities outlined in chapter 5 (Basic Defense), remembering that any degree of success here might only delay the inevitable. That would be the story for any part of the galaxy where a contest between two such unequal combatants was in progress or pending.

If we should have attained a measure of interstellar capability by the time an attack on us is mounted, the respective technologies could be regarded as more equal. Accordingly, we might be able to counter a space enemy's sophisticated weaponry and retaliate to such an extent that he broke off the engagement permanently and went on to look for more lucrative cosmic pastures. This is really only an extension of the balance of power on this planet at the present time.

Having been wantonly attacked, we could even carry the attack to the enemy's celestial shores. If his aim were to gain an armed foothold in the Solar System, it would hardly be unethical if we dispatched a task force with a few "presents" to his source of origin to illustrate graphically to him the error of his ways and the inadvisability of any further adventures along the same lines. News of the event over a galactic communications system might also tend to discourage other incorrigibles. To carry a cosmic war to an enemy's home star system and planet assumes of course that we are conversant with their location and, moreover, that our interstellar capability has as great a range as his. Obviously if he can span 20 light-years while we are restricted to 5 his home system is sacrosanct.

It might be of interest at this stage to put ourselves in the position of aliens from planets of an expanding, colonizing front who are considering future operations with respect to the Solar System. What are their tactics going to be? We could reasonably assume that they are aware by virtue of our radio transmissions that life and a civilization of sorts exists on the third planet outward from the Sun. In their eyes it will not be a very wonderful civilization. Nevertheless, they could be aware that it did possess nuclear arsenals and had a limited but growing space capability. They would also realize that as yet it had virtually no hegemony over the star system in which it found itself. The aliens have therefore little doubt about their capacity to take over Earth. For the moment, however, there is no particular haste. Though terrestrial civilization could be overcome, it could still prove a little troublesome. A policy of continued surveillance might be regarded as the most satisfactory during the preparatory period.

At this point we will make another assumption. The aliens happen to be members of a great galactic federation, which would prefer including and integrating the terrestrial civilization peaceably. They see this as best in the long run for the peoples of Earth. The vanguard of the galactic federation force is therefore instructed to carry out a policy of watch and wait. But to watch means rather more than sitting out well beyond Earth's orbit trying to understand the vituperation and saber-rattling emanating from so many of Earth's so-called leaders. Real surveillance means precisely what the words imply—that the affairs of Earth should be monitored from a point as close to that planet as possible. It will be impossible, they realize, to conceal such operations, but in veiw of their highly sophisticated technology this may not matter a great deal. Their ships can outmaneuver the crude clumsy spacecraft and aircraft of Earth to such an extent that to the inhabitants of Earth it is a case of "now you see us, now you don't!" We might imagine these beings from the stars saying: "These people are still quite primitive. When they see strange objects in their skies they will not know what to think. Those who don't see them won't believe those who do. And if occasionally we happen to show ourselves as living beings during brief excursions on the surface of their planet, the few who see us will be disbelieved and ridiculed."

This situation is admittedly contrived as a possible explanation for UFOs and the peculiar way in which they appear to operate. Whether or not UFOs are real and originate from remote worlds of other solar systems we still cannot say. Nevertheless, the scenario just described does fit the circumstances. It would, in the light of present evidence, be unwise to disregard all reports of UFO sightings. As a race it might be a good idea from time to time to disengage our minds from trivia and look up!

Finally, we must consider the question of tactics in its supreme (and probably most sci-fi) form—the bold and brazen entry into our Solar System of a fleet of alien starships hell-bent on trouble and their interception by terrestrial craft more or less their equal. Here we have the classic battle of the giants— combat between a ruthless, advanced, alien civilization and an equally advanced terrestrial society of the future. Precise tactics and maneuvers are difficult to envisage. The basic necessity for any defending force is to destroy the enemy—which entails the destruction of his fighting craft before his transports can unleash the actual invading force. The destruction of the latter can then be carried out at will. In such a match superior armament, efficiency, and speed of the flagships would go far toward determining the final result. Due allowance must also be made for darting attacks by ultrafast, highly maneuverable small craft. Several such attacks well pressed home could in a running battle inflict quite unacceptable losses on an enemy space fleet.

We cannot rule out the kamikaze type of attack either, and if the tide of battle were running against the weaker side, that could well become part of the strategy just as it did in Pacific waters during the years 1944 and 1945. The deliberate crashing of a small explosive-laden spacecraft in the vitals of a large hostile starship or troop-carrying space freighter would, we must assume, prove even more disastrous to the latter than the Japanese suicide fliers were to the aircraft carriers of the United States Pacific Fleet toward the close of World War II.

We might go even further back into history by invoking the concept of Drake's fireships. During the attack by the Spanish Armada against England in 1588, a ploy of Sir Francis Drake, the English admiral, was to send several blazing ships sailing out of control among the closely assembled ranks of Spanish galleons and troopships. We can hardly compare sophisticated alien starships with the cumbersome galleons of the past. The principle, however, could easily be enshrined. As the hostile alien space fleet draws close to a large asteroid or moon, a host of unmanned obsolete craft containing massive nuclear charges is directed into the fleet's midst and detonated by impact or remote control.

The question of tactics, tricks, and subterfuges must remain a very open one. We feel bound to repeat, that although we have chosen to involve our own planet and our own kind in most of the hypothetical possibilities, we have done so largely to bring home to readers the realities and enormities of interstellar war. All that has been said in this chapter, and indeed in this book, could actually be taking place *at this very moment* deep among the stars of our galaxy. The theme of space weapons and of space war is not one that applies only to this particular corner of the Milky Way. It could be universal in the truest sense of the word. Perhaps if our technology were more developed, perhaps if our sensors could penetrate more deeply and more sensitively into the mass of stars that constitutes our island universe, evidence of mortal cosmic combats might be forthcoming. Evidence of the existence of benign, socially conscious, galactic federations could also be forthcoming. Both seem equally plausible. Evidence of the latter could arouse great visions and even greater hopes; evidence of the other could serve as the awful warning!

17. ARE THEY HERE ALREADY?

We have not yet touched on one aspect of our theme—summed up succinctly by the title of this chapter. Whether or not we should do so may be considered by some to be a highly debatable point.

We have considered the approach of aliens and the arrival of aliens. We have dwelt on the possible forms of attack and come to the unhappy conclusion that most would be decidedly unpleasant—unless *we* one day happened to be the aliens looking for new worlds to conquer. We have also said a little on the subject of defense. It may be thought that the question of defense should have been given more attention. But let us be quite frank here. An attack by intelligent beings from planets light years out in the galaxy, beings perhaps thousands of years our seniors, is going to prove almost impossible to counter—at least for any length of time.

Let us look on it this way. In a relatively few years now manned expeditions from Earth are going to descend upon the planet Mars. As things stand at the moment it looks as though the planet is void of life, although it is impossible as yet to be absolutely certain. The two recent American *Viking* landers have revealed from two separate points on the surface of the legendary red planet only a desolate waste of rocks and low undulations. No life on Mars then? So certainly it would appear, though an old proverb here on Earth might just be applicable in the circumstances: "Two swallows do not a summer make!"

It should always be borne in mind that two probes sent by Martians to Earth, both landing in disparate desert regions, would not reveal the fact that Earth teems with intelligent life.

Let us suppose, however, that our first manned expedition to Mars *has* revealed the positive presence of advanced life, advanced, that is, in the strict biological sense. These Martians are not advanced in the technological sense (for which reason they may be very much happier). We would term them primitive, since their only weapons are spears, clubs, and knives. The bow and arrow, their ultimate weapon, has just been invented. The Martians take strong objection to our landing on their hitherto sacrosanct world and make their displeasure known by fierce attacks on our people, who suffer several casualties, some fatal. When this intelligence reaches Earth it does not go down too well, and almost immediately an armed force is dispatched to Mars. The rest follows as night does the day. The invaders may be handicapped by working in a different gravity and having to use life-support suits because of the very low oxygen content of the Martian atmosphere; nevertheless armed with grenades, mortars, and high velocity automatic weapons, they are more than a match for the primitive but biologically advanced inhabitants of Mars. Despite the desperate attacks of the Martians the conflict becomes a glorified "turkey shoot."

The degree of difference between these hypothetical Martians and our own kind could, technologically speaking, be likened to that existing between advanced aliens and ourselves. In fact, the difference between the aliens' hardware and techniques and our own would probably be very much greater—and *we* would be the underdog.

To talk of defense against attack by a civilization so vastly superior is really a case of living in cloud-cuckoo land. In the circumstances we would have to do what we could, but little in the way of success could reasonably be expected. Not very much can be said, therefore, on the subject of defense, which is why it has largely been ignored in these pages. In science fiction, Earth and its peoples nearly always secure a marvelous last-minute reprieve. This would be highly unlikely in the actual event.

And so to the theme of this chapter: Is it possible, however remotely, that the vanguard or fifth column of an alien invasion force from the stars is with us already? At this point there could be readers tempted to say, "Oh, not again, not more UFO's!" or something along those lines. This is quite understandable. There has been a plethora of "flying saucer" books in recent years. It is not the present author's intention, however, to delve into these matters here. They *have* a degree of importance, perhaps greater than many of us imagine or care to admit, but as a subject they have been well explored already.

Despite the tremendous number of reports of unidentified flying objects, we have, as yet, no clear, categorical evidence that these emanate from other planets and contain living, intelligent beings. This is not to say or suggest that such a thing is impossible—only that, so far, *no* unequivocal evidence to substantiate the claim has been forthcoming. Indeed, it is also true to say that no unequivocal evidence for the actual *existence* of UFOs has yet come to hand. In view of the continuing reports of sightings, however, we must assume that, whatever the reason, something odd *is* happening out there. All those who report sightings cannot be knaves or fools.

If UFOs represent alien spacecraft, then they are assuredly of a very peculiar kind and certainly do not *appear* to be inimical to us. This could so easily be an illusion. A reconnaissance mission is supposed to do just that—to reconnoiter, not to attack. At the same time reconnaissance is supposed to comprise discreet, intelligence-gathering surveillance. The antics of UFOs could hardly be termed discreet for, according to so many observers, they cavort about the skies in the most unashamedly flamboyant manner. Indeed since the inception of the UFO busines back in 1947, this flamboyancy has been a very notable feature. Some of the most recent reports confirm the same tendency. For example in May 1977, in Dereham, Norfolk, England, several people reported seeing "a dome-shaped object with portholes and a pulsating red light underneath." The same object was seen from the county of Buckingham farther to the west by a young couple who described it as "a Mexican hat with dome and flashing light, gliding *silently* across the sky." In September of 1977, in the southeast county of Kent, a young medical secretary, Dianne

Howard, reported a very similar sort of object in the sky. According to Dianne, the rim lights were "numerous, red and
green and there was a significant humming noise." It could
hardly be said that the reports fit in with the concept of discreet
surveillance—unless of course the aliens (assuming these are
aliens) regard us as so stupid, primitive, and ignorant that surveillance doesn't need to be discreet. But then if we are all those
things, why bother about surveillance at all?

If UFOs have extraterrestrial and extrasolar origins, they
must nevertheless be viewed as surveillance craft probably emanating from a "mother" starship. Where, it can be asked, is this
"mother" ship likely to lie? That is a question to which we will
return in a moment or two.

The query concerning the presence of a "mother" ship assumes that UFOs are not in themselves capable of making the
journey between their sun and ours. We should not, of course, be
unmindful of the possibility that they might represent the futuristic creations of a civilization so advanced that transit through
some strange hyperdimension has been rendered commonplace.
The basic essentials underlying this particular concept were
explored at some length in the forerunner to this book, *Interstellar
Travel: Past, Present, and Future,* to which the interested reader is
referred. But so vast is interstellar space that presumably even
great "mother" starships must adopt some such technique if
they are to negotiate the galaxy with any measure of freedom.

So we return to the question of the precise whereabouts of the
parent vessel responsible for the swarms of UFO probes. Such a
vessel, we would assume, must be of considerable dimensions.
For this reason alone it is hardly likely to cavort around within
or just above Earth's atmosphere. Moreover, if mere surveillance
is, for the present, the name of the game, this will be left to its
UFO offspring, whose antics and significance the "ignorant"
terrestrials do not comprehend. Its own mighty presence would
lead even the most "ignorant" terrestrials to certain immediate
and inevitable conclusions. So, to escape visual and radar detection, it retreats well out into the darkness, taking up a parking
position well beyond the moon or even farther.

Could "mother" ships have approached much closer to this

planet before the era of terrestrial radar? The only possible
answer is that if alien starships entered the Solar System then,
they would soon have realized that neither radio nor radar
radiation emanated from the Earth. In the circumstances their
occupants could have concluded that it was safe to stand in
much closer to our world without fear of detection. Is it possible
that from time to time they even entered the upper layers of our
atmosphere? Presumably so. In the circumstances, would there
have been any pronounced sonic or other physical effects? Such
effects are certainly not impossible, and a number of deep booming sounds, apparently emanating from the upper atmosphere,
were reported over a number of years during the last century. It
is true to say that these have never been adequately explained.
This aspect was also dealt with in the pages of *Interstellar Travel*
and for that reason will not be treated again here other than to
update the position. Updating is both appropriate and necessary
since reports during the early months of 1978 spoke of a series of
"atmospheric booms" that rattled windows and tripped scientific instruments that measure air pressures along a wide swath
of the east coast of the United States from Charleston, South
Carolina, to the Canadian province of Nova Scotia. On some
occasions flashes of light were said to have accompanied these
booming sounds. This latest outburst started on December 2,
1977, when two exceedingly loud booms were both heard and
felt in the coastal city of Charleston. Residents along the New
Jersey coast to the north had a similar experience later that
same afternoon. On December 15, Charleston was again reputed
to have been "rocked" by five very loud and inexplicable booms.
At the same time explosions were also heard by independent
observers off the coast of Nova Scotia. Come December 20, two
further booms were clearly heard overhead in Charleston once
again. New Jersey observers reported a single detonation. More
followed over different locations on December 22 and December
30, and on January 5, January 12, and January 18, 1978.

Explanations for these strange sounds have ranged from supersonic aircraft to the ignition of methane gas pockets by static
electricity. Neither explanation is particularly credible and, in
fact, no conclusive evidence has so far emerged to support an

orthodox explanation for these mysterious and, at times, rather alarming sounds. The United States Department of Defense, along with other government agencies, denied responsibility for any connection with these sounds. With the denial of clandestine experiments or the like, the way has, perhaps unwillingly, been cleared for speculation. One of the really odd features is the evidence of individual booms and rumblings being heard over extensive reaches of this particular coast. In this connection it must be added that residents in the Connecticut towns of East Haddam and Moodus have reported strange rumblings high in the sky on odd occasions since *1829*, and inhabitants of South Carolina have frequently heard and reported peculiar and inexplicable offshore noises. The matter is now said to be in the hands of the Naval Research Laboratory, which has been requested to investigate and report on the mysterious sounds. This could prove a rather difficult task. Already, attempts are being made to attribute the sounds to the sonic booms of distant jet aircraft, carried over a very great distance by unusual refractive effects in the atmosphere. Inevitably, this explanation carries a ring of truth. Unfortunately for this theory, we are forced to point to the incidence of the phenomenon years *before* the advent of jet aircraft.

A particular instance stands out very clearly in the mind of the author from boyhood days, though at that time thoughts of aliens, UFOs and spacecraft were nonexistent. The year was 1936, the precise date July 19. The region concerned was near the small Scottish border town of Peebles, which lies about thirty miles to the south of Edinburgh. The town nestles charmingly among the pine-clad slopes of the Moorfoot Hills. At that time the writer had become a rather enthusiastic amateur geologist (collector of rocks might be a more appropriate if less grandiose title). Collecting entailed frequent expeditions along the hill tracks to quarries and rock outcroppings. The day in question was warm and sunny with no hint of rain or adverse weather of any kind. When several miles from the town itself, by the side of a stream from which at times (allegedly) small traces of gold-bearing quartz could be found, there came very clearly on two occasions about five minutes apart two distinct, distant rumbling sounds which, whether they did or not, certainly seemed to emanate from above. Now Peebles, because of its position among hills, is noted, especially at that time of year, for its not infrequent and sometimes rather violent thunderstorms. Such storms, however, are almost invariably preceded by a still-ness, lack of wind, heavy sultriness, and swiftly darkening skies. On this occasion all these features were totally absent. The day was clear—and it remained clear and, as could be determined afterward, there were no electrical storms in the vicinity. Being miles from the town and from any shelter, the idea of a sudden downpour and the flash floods that such a downpour could cause among these little meandering hill streams was none too comforting. Yet never did conditions less indicate an approach-ing thunderstorm. Occasionally blasting operations took place in some of the nearby quarries, but the sounds they made were totally different and quite unmistakable. If thunder was out, so were quarrying operations. Thus the reason for the rumblings, which were a most odd sound (and to a schoolboy mind rather sinister) remain inexplicable. It must be added at this point that they were also heard by others in the town and district. Presum-ably, therefore, imagination can be ruled out.

Oddly enough, this was not the end of the story by any means. At that time a comet had made its appearance in our skies—Comet Peltier, of 1936. Though not a brilliant object, it was discernible if one knew precisely where to look. The evening of that same day, around midnight, while locating Comet Pel-tier in a clear, moonless and star-strewn sky, these same odd rumblings could once again be distinctly heard. What renders all this so unique is that several people in the area reported seeing a "cigar-shaped object emitting a bluish light" passing across the night sky at that time. The writer must admit that, although he picked out Comet Peltier among the stars, he did not see the other peculiar object which may, of course, have passed across the sky in a different direction from that in which the comet lay. Certainly there can be no question that the comet and the other mysterious object could be confused. All this, it must be emphasized, took place in July 1936, long *before* the advent of jets or the present UFO era.

A past account of similar rumbling sounds has just come to the writer's attention. It is in the form of a letter written to the scientific journal *Nature* by G. H. Darwin, son of the famous Charles Darwin. The letter appears in the issue of October 31, 1895, and contains a number of highly interesting points. Darwin states that his attention was first drawn to the subject by the Conservator of the Museum of Natural History of Belgium, one M. van der Broeck, who spoke of "curious aerial or subterranean detonations which are pretty commonly heard, at least in Belgium and in the north of France, and which are doubtless a general phenomenon, although little known, because most people wrongly imagine it to be the sound of distant artillery. I have constantly noticed these sounds in the plain of Limburg since 1880 and my colleague of the Geological Survey, M. Rutot, has heard them very frequently along the Belgian coast, where our sailors call them 'mist pouffers' or fog dissipators. The keeper of the lighthouse at Ostend has heard these noises for several years past; they are known near Boulogne and more than ten of my personal acquaintances have observed the fact. The detonations are dull and distant and are repeated a dozen times or more at irregular intervals. They are usually heard in the daytime when the sky is clear and especially towards evening after a very hot day. The noise does not at all resemble artillery, blasting in mines or the growling of distant thunder." M. van der Broeck then goes on to attribute the sounds to "some peculiar kind of discharge of atmospheric electricity" which, in that day and age, was a reasonable sort of statement.

As though booming and rumbling sounds in the skies were not a sufficient mystery, reports of "humming sounds" have recently come from people in a number of British cities. These are not, it is said, in any way reminiscent of the sound of distant aircraft, either jet or piston. Similar sounds are reported to have been heard a century ago off the coast of Greytown, USA. The latter were generally heard by persons on board iron steamers and the phenomenon was accordingly attributed in some way to the hull of the ship. If the "humming" sounds heard in British cities at times today is identical, then ships' iron hulls can hardly be blamed.

Now it would be all too easy to claim airily that these sounds, past and present, originate as a consequence of large, fast star-ships maneuvering in the upper reaches of our turbulent at-mosphere—or even of UFOs performing their inexplicable gy-rations. It must be freely conceded, however, that there exists no more proof for this than for any other explanation. It is *only* a possibility, no more and no less, as yet. The sounds are undoubt-edly very real and have been heard on and off over a consider-able period of time. Similar sounds have been reported from various other parts of the world and in many cases they predate, by a considerable margin, the advent of aviation. All reports seem to indicate that the sounds emanate from *above* though, of course, loud detonations from a source lower down or even on the surface of our planet could be directed upwards and then refracted downwards if atmospheric conditions happened to be appropriate. In this way they could be heard at a considerable distance and over a considerable area, though hardly, one would have thought, in such strength. Many will inevitably cite the degree of highly sophisticated radar surveillance now exercised over our world by virtue of which any alien craft ought at once to be detected. They have a very valid point and its truth cannot be denied. Our alien visitors could, of course, have some very clever tricks up their sleeves in this respect. During the last war we "blinded" German radar by the simple expedient of drop-ping "window," masses of tinfoil. An alien civilization could have its expedients, too. Certainly UFOs rarely seem to suffer radar detection. There, for the present at least, the continuing saga of mysterious booms from the sky must remain.

The next question is whether or not there is evidence to substantiate the belief that alien craft of one sort or another are lying "offshore" so to speak, i.e., lurking somewhere in the dark-ness well beyond the Moon. The answer to this is presently no, which is not the same as saying that no such craft are there—or ever will be. So far our surveillance of these regions has been minimal. This will become less and less true as the solar system is explored more extensively.

In 1973 an interesting claim was made to the effect that a spacecraft emanating from a planet of the star Epsilon Boötes had, in fact, arrived within the confines of the Solar System and was circling the Earth in an equilateral configuration with Earth and the moon (Fig. 8). The claim specified that the craft was a *space probe* and not a starship. This is understandable since the object is supposed to have arrived in the Solar System 13,000 years ago and lain silent until the advent of terrestrial radio set its electronic circuits functioning. There may, at first glance, seem less menace from a space *probe,* especially one arriving so long ago, than from a huge starship perhaps full of alien shock troops equipped with sophisticated weapons of ultimate destruction. This is true, but only so far as it goes. If such a probe does lie out there, it proves at least and all too plainly the vulnerability of the Solar System to spacecraft from beyond!

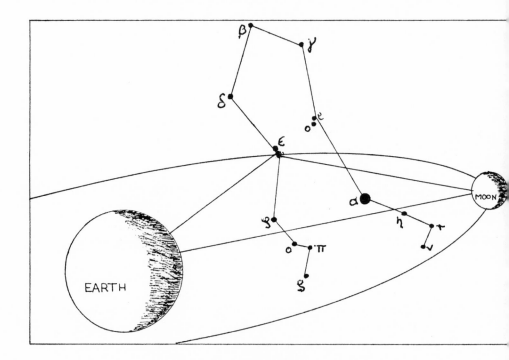

Figure 8
Position of possible alien probe (ringed)

The claim rests on a new interpretation of certain delayed echoes of radio signals transmitted by a terrestrial radio station, PCJJ, Eindhoven, Holland in 1927–28. The basis of the concept must be attributed to R. N. Bracewell of the Radio Astronomy Institute, Stanford University, in a paper published in 1960. In this paper, Bracewell, an eminent radio astronomer, suggested that if advanced communities were spread throughout the galaxy at distances upwards of 100 light-years, unmanned space probes might be the most effective means of communication between them. On reaching and entering the Solar System such a probe might monitor our radio transmissions and then re-transmit some of them back to us. The retransmitted signals would thus appear as "echoes" of the original with delays of several seconds or minutes such as those reported from station PCJJ, Eindhoven.

It must be admitted that if this is an alien method of telling us something it seems a peculiarly illogical one. But then what might so easily be illogical to us could be thoroughly logical to alien minds. During the late twenties, when these "echoes" were first heard, little attempt was made to do anything about them since the phenomenon was understandably regarded as a natural one. Several experimenters did, however, try a graphical analysis by plotting the delay time of a particular echo against its position in the sequence. Pulse sequence number was made the "X" (horizontal) axis of the graph and pulse echo delay time the "Y" (vertical) axis. The end product of this intellectual endeavor was precisely nothing. In 1973, D. A. Lunan decided to reverse the two axes. When he did this a striking though incomplete resemblance to the constellation Boötes, the Herds-man, resulted (Fig. 9). On this graph one star is missing and another is misplaced. The missing star is Epsilon Boötes. Lunan's theory is that this star represents the origin star of the probe and claims that if we had retransmitted the graphical "map" to it with Epsilon Boötes inserted in its rightful place, the probe's full contact program would have been initiated. The incorrectly placed star in the graph is Arcturus (Alpha Boötes), the brightest star in the constellation. On the graphical representation of the constellation the star appears well above and to the left of its true position. Could there possibly be a valid reason for this? One almost immediately suggested itself. Arcturus has one of the largest known proper motions, 2.29 seconds of arc per year—which is equivalent to the apparent

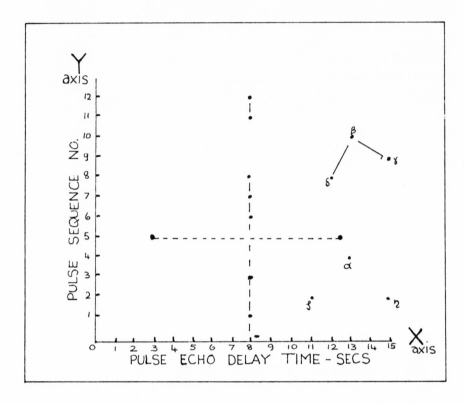

Figure 9

diameter of the full Moon in 800 years. This motion is toward the southwest. On this basis the explanation might be as follows: the probe arrived in the Solar System from a planet of the star Epsilon Boötes 30,000 years ago, compiling its star maps *at that time*. This part of its mission completed, it became quiescent until reactivated by ionospheric reflection test transmissions by ourselves.

There is assuredly a very high measure of speculation here and still it all seems remotely feasible. One factor in favor of the theory is the almost uncannily accurate representation of the constellation Boötes that resulted from graphing with reversed axes. If it is entirely due to coincidence, then that coincidence can only be described as remarkable in the extreme. And from what we now know of Arcturus' proper motion, its position on the graph could well have been its position when the probe arrived in the Solar System (Fig. 10).

Lunan then proceeded to extrapolate his basic tenet to a very considerable extent. This, perhaps not surprisingly, led to criticism that he might be endeavoring to equate fact with theory.

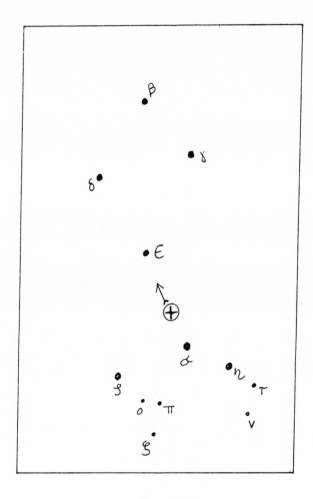

Figure 10

He was at least trying to present a reasoned case and his attempt should be scrutinized and not just arbitrarily torpedoed because it deviates from the conventional. Perhaps the most valid argument against the extrapolations is that they do seem, especially at a first glance, to be extraordinarily complex. The complexity leads to doubts, of course, since an alien civilization would certainly have devised some more straightforward means of imparting the intelligence it wished to convey. This argument in turn can be countered by the argument that, despite our beliefs to the contrary, we are not very bright by galactic standards (which is probably not far removed from the truth).

The conclusions accruing from these extrapolations were as follows:

1. We [the aliens] come from the star you know as Epsilon Boötes which is a double star.
2. We live on the sixth planet of seven, counting outwards from our sun.
3. Our planet has one moon: the fourth planet has three. The first and third planets have one each.
4. Our probe is in the orbit of your Moon.
5. The position of Arcturus in our maps should be updated.

Attempts to make contact with the probe by radio have been made, so far without success. In this respect the writer has been associated with Lunan in a minor way.

The question (enigma might be a better term) of odd echoes and even snatches of long-past voice and Morse transmissions has recently aroused considerable interest. So far as transmissions are concerned, the possibility of hoaxes must always be borne in mind, for this is an area of activity in which the lunatic fringe loves to operate. Just why is known only to them. There have been instances, however, in which the circumstances have tended to rule out the chances of a practical joke.

The writer (a licensed radio amateur) has to date only experienced one of these odd transmissions. It comprised a series of messages in Morse picked up on the 144 MHz amateur VHF band one evening in August 1977. The first purported to come from the giant British airship R101 shortly before it crashed and burst into flames near Beauvais, in northern France on October 5, 1930, while on its maiden voyage from Cardington, England, to India. It was followed shortly afterward by another, apparently from the French airfield at Le Bourget near Paris, to this effect: *"G-FAAW a pris feu. C'est une grande catastrophe!"* (G-FAAW has caught fire. It is a major disaster!) G-FAAW were the registration letters of R101. Later examination of inquiry records and newspaper accounts of the time showed the messages to be *precisely* those received from the airship and from Le Bourget.

The tone of the Morse note indicated that the signals were emanating from the type of transmitter then in vogue but long since outmoded. The actual messages from the airship and from the airfield were *not* recorded (at least on Earth) in 1930. It should also be added that they went out originally on the low frequency (long wave) band. When heard again in August 1977, they were being received on VHF, almost as if a recording of the original were being transmitted.

What is one to make of an event of this nature? At the time it was a rather unnerving experience with strong undertones of the supernatural. Here was a message coming from an airship flying in 1930 to the effect that all was well. An operator in 1977 knew only too well that this was not so and that soon the great ship of the skies would be a flaming funeral pyre on a French hillside. The temptation, a ridiculous but very strong one, was to transmit a message in return, warning the huge dirigible that it was flying to its doom! There is a peculiar popular notion that radio signals go on forever and can perhaps some day be picked up again. That is, of course, not the case.

The writer wishes to make it *very clear* that he is *not* categorically claiming that these "echoes" of nearly half a century ago emanated from some mysterious alien probe lurking somewhere near the Moon. As a practical joke it would have been in the worst possible taste since almost the entire complement of passengers and crew perished in the dreadful holocaust when five and a half million cubic feet of highly inflammable hydrogen ignited. But as a practical joke it would have been rather difficult to carry out. It can also be recorded that the identical messages were picked up by another radio amateur in Hull, England, at the same time. The intriguing point about this is that a hoax VHF signal, because of its relatively short range over land, should not have been received in western Scotland and in east central England at the same time. Another highly intriguing (or should we say disturbing) feature is that the bearing of the beam antenna of the writer's station was, at the time, directed toward the point on the horizon that coincided closely with the supposed celestial position of the Epsilon Boötes probe. This highly pertinent fact became apparent only later.

Was this a genuine synthetic echo originating from somewhere in the darkness beyond our planet, an example of the supernatural or a stupid but skillful hoax? The truth may never be known. As we have said, a hoax along these lines would be difficult to accomplish and the writer is certainly not a strong believer in the occult. That leaves only the idea that something, somehow, and somewhere picked up, recorded, then retransmitted on VHF nearly half a century later the messages from a doomed airship. And for obvious enough reasons that is still a very difficult proposition to accept. The entire question of odd radio "echoes" and even odder radio transmission replays must, it seems, remain open.

We may appear to be digressing somewhat from our main theme. Surely mere probes represent no menace to us. Here it must be said that much would depend on the nature, origins, and purpose of that probe. An apparently friendly probe programmed to give information relating to its home planet and civilization in exchange for the same about our own could be seeking that information for sinister reasons. And the mere fact that a space vehicle, albeit an unmanned and unarmed probe, had made the journey from the environs of another sun to those of our own should be food for very serious reflection. If that probe from Epsilon Boötes really exists, it is not one that ought to worry us, since its arrival took place so very long ago. But any lurking silently in the Solar System unknown to us, sending vital coded intelligence to an aggressive, acquisitive civilization, would represent a very considerable source of danger.

From time to time we are treated, generally by the popular press, to reports that would indicate that the vanguard of the alien host is already with us. Entire volumes have appeared whose authors claim, for example, direct contacts with aliens or flights in their craft. Such books are best treated as the science fiction they undoubtedly are.

Reports that tell of the actual appearance of supposedly alien beings *must* be treated with the greatest possible reserve. Some of these fall into the same category as the aforementioned books. Others must be attributed to hysteria or highly vivid imaginations. A few, *and only a few,* are worthy of much more careful scrutiny and analysis. Undoubtedly there are instances where perfectly sincere, rational, and responsible people have had some extremely frightening and inexplicable experiences. *If* (and this word must be stressed) aliens have from time to time set foot upon our planetary shores, then a highly disturbing state of affairs exists, for it would indicate that, depsite all our sophisticated techniques of electronic surveillance, they can get through undetected and with consummate ease. Moreover, in almost every case, the supposedly alien beings do not appear to have been particularly friendly or understanding. It must be emphasized, however, that no unequivocal instance of an alien presence on the surface of our planet has ever been established and these lines are in no sense meant to indicate one.

The entire UFO question might be less worthy of attention had there not been such a plethora of reports. What could be a highly significant feature is the manner in which so many of these reports correspond. Equally significant is the undoubted fact that many come, not from the lunatic fringe, the neurotic or the easily excited, but from highly responsible people trained by profession to observe and report accurately. It is suspected also that many others have decided on sealed lips rather than expose themselves to ridicule. The writer feels that the final words in this context must, and can only be: WATCH THE SKIES!

* * *

Shortly after this chapter was written UFO sightings were reported only a few miles from the writer's home, but regrettably he was not sky-watching that particular evening. The date was Tuesday, April 18, 1978. The object, phenomenon, or whatever, was seen by several people in the little Ayrshire village of Dunlop. By common consent it was silent, round, orange red, very bright, and quite definitely spherical. In the first instance it was seen by two Scottish police officers as it traveled in a westerly direction over the nearby village of Stewarton. Said one of the officers later: "It was curious—too bright for a helicopter, too low for a conventional aircraft. I thought it might be a flare but flares don't travel so fast or so straight."

A farmer and his son a few miles away reported the sighting independently. Their description matched that given by the police officers, except that on this occasion the thing appeared to alight on a field. By the time they reached the field it had apparently gone. Mystified, the pair returned to their farm. Thirty minutes later they clearly heard from the general direction of the same field what they described as a screeching, bloodcurdling sound, as if from a mortally wounded animal. A neighboring farmer and his wife also saw the strange red light as its beams shone through a glass door. When they looked out they could see the object plainly, still and silent on, or just above, a field between their farm and that of the other farmer. Their comment was: "It sent a shiver up our spines. We have never before seen anything like it." These reports are documented and appeared in the *Glasgow Herald* of Tuesday, May 2, 1978. Whatever it was, these people, good solid down-to-earth citizens most assuredly saw *something!* In the normal course of their lives it is at best doubtful if they would even have thought of such things as unidentified flying objects. There can be no question either of ball lightning or any other related phenomena. In the first place, ball lightning would not behave in this fashion. In the second, checks made with the appropriate meteorological authorities indicated no trace of thundery conditions that evening. The season for electrical storms in southwest Scotland is from June to September.

It should be added that on the previous night there had been many reported sightings in England as well.

A paper by D. Herbison-Evans of the University of Sydney, Australia appearing in the *Journal of the Royal Astronomical Society* of London for 1977 may be of interest as a final footnote to this chapter, emanating as it does from a highly reputable quarter.

Australia is a country from which we do not normally hear very much on the subject of UFOs. The author of this paper points out, however, that 20 percent of Australian sightings are truly unidentified flying objects inasmuch as they are not attributable to hoaxes, imagination, or misinterpretation. Of these true UFOs, 65 percent were witnessed by more than one person and about 10 percent come from experienced and highly reputable observers such as police officers and meteorologists.

On a worldwide basis sixty-eight UFO sightings have, in fact, been reported by astronomers.

The author of the paper remarks: "Bearing in mind that it requires only *one* authentic report of an extraterrestrial visitation to refute the negative assumption, it is only necessary to consider these sightings which have high credibility and high strangeness." In this context he relates three sightings which, after an extensive survey of the literature, he believes to be the most significant. These are:

1. An object plainly seen by air traffice controllers and tracked on radar at Kirkland U.S. Air Force Base, New Mexico, in 1957.
2. A sighting in Papua by the Reverend B. Gill and some friends in 1959.
3. An incident in 1964 in which a young woman and her husband claim to have been abducted by an alien spacecraft. Normally this is the kind of story that should be looked on with a very jaundiced eye but for the fact that the woman was able afterwards to reproduce a rough copy of a chart that she said she saw within the craft. This chart purported to show twelve "trade route" stars and twenty background stars. It was found subsequently that a projection of the three-dimensional positions of our twelve nearest single dwarf stars in the spectral range F6 to K1 fitted the so called "trade

route" stars drawn by the woman remarkably well. Indeed, on the same projection other stars appear that correspond remarkably well also to the woman's background stars.

In conclusion Herbison-Evans states: "Thus there is evidence that we are being visited but the evidence is poor. The main problem with UFO's is that, like quarks and total solar eclipses before they were understood, they are difficult to make into reproducible observations."

18. DANGER SPOTS

So far, we have given no thought at all to what is undoubtedly a most important aspect—the particular stars from which an invasion of our world might originate.

Despite the great profusion of stars within our galaxy, it is only those falling within certain categories that are likely to have planets capable of spawning life. An even stricter selection designates those where life has the requisite intellectual capacity to initiate and develop interstellar travel. We must not, therefore, fall into the trap (admittedly an easy one) of looking up into the night sky and arbitrarily assuming that every star we see has planets and that on all these planets there are intelligent advanced beings capable of reaching our world. This is not so, and never will be.

Present astrophysical reasoning based on the ages of stars and their speeds of rotation indicates that about 67 percent of all the stars in our galaxy could have planetary systems. Obviously this implies the existence of a tremendous number of planets. Taking the Solar System as a standard (and there are sound reasons for doing so), only one planet in nine, our own, has, to the best of our knowledge, spawned life. Certainly it is the only one to have produced intelligent life—unless any artifacts ever found on Mars indicate a civilization there, which eventually perished. This means that of the great number of planets in existence in our galaxy roughly 90 percent are probably unsuitable as life-initiating and developing centers, leaving only a meager 10

189

percent that could be. We should really say "apparently meager," for 10 percent of a vast total constitutes a by no means
inconsiderable number in itself. As a notable mathematician
once said, "1 percent of infinity is still infinity!"

By going into the elementary astrophysics of the matter, we
should therefore be able to list a number of stars, some of them
clearly visible in our night sky, that just might conceivably
represent a source of mortal danger to our world, its peoples,
and the civilization we have created. This we will presently do.

Another important point must also be stressed at this juncture. As we gaze up into these very lovely stellar depths, we
should, provided the moon is absent, be able to discern the
Milky Way, that great gown of stardust, which night trails
across the heavens. This represents the galaxy proper, the vast
preponderance of the great island universe to which we belong.
The many bright (and not so bright) individual stars we see are
merely those lying relatively close to the Sun, even though
distances of several hundreds of light-years can often be involved. The diffused band of light we call the Milky Way is, in
actuality, a tremendous accumulation of stars so remote that the
unaided eye discerns their cumulative light only as a faint amor-
phous ribbon, girdling the night sky.

Now if interstellar travel by unique and specialized means is
feasible, i.e., hyperspace dimensions or black holes, then danger
could threaten us also from these incredibly remote regions,
remembering that extradimensional travel would probably involve near instant transit. If, therefore, a visit to our planet by
aliens ever occurs (for whatever purpose) it does *not* auto-
matically follow that their place of origin is one of the "close"
appropriate stars in the list we will presently draw up. They
might well have come from a star system out on the far rim of
the galaxy. Clearly, we cannot draw up a list detailing every star
in the galaxy around which intelligent life could have developed. Consequently, we will look at the stars of the galaxy
generally and assess, so far as contemporary human knowledge
will permit, the potential of various classes in an exobiological
role. In closing, however, we will try to be more specific regarding the so-called "close" stars that we can see every clear night.

There is a tendency in postulating other possible galactic life forms to speculate on noncarbon (e.g., silicon-based) life. It is a fact, however, that unless the laws of chemistry, physics, and biology are miraculously suspended elsewhere, no chemical element offers even remotely as stable a base for living matter, especially intelligent life, as does carbon. It is the most likely element and therefore the one on which our considerations and premises will be based. This does *not*, of course, rule out others entirely. It is very difficult, however, to comment in any depth on hypothetical life forms based on alternative chemistries.

Biologists believe, with good reason, that life on Earth would have been a very transient thing (possibly existing for only around 50 million years) had that elementary life not been able to link its processes to the pervading energy of sunlight. Thus, the star we call the Sun not only nurtures life today, it also made possible the development of that life 3,000 million years ago. The central sun of any planetary system is therefore of vital importance in respect to any future biology in its environs. The principal parameters involved are size, temperature, and age, all three being interrelated. These link the development of life with the star and form the essential bases for pinpointing the most probable whereabouts of intelligent life forms elsewhere in the galaxy.

It also seems more rational to base our considerations on terrestrial environments only, although in this context the word *terrestrial* must be given less sharply defined limits. Over the years writers of science fiction have had a wonderful time postulating intelligent life in the most unlikely spots. About this we should not complain, for these books and stories were meant to entertain, not to instruct, and that they have done fully and well. Fiction, however, belongs only within the pages of fiction. We cannot, therefore, envisage Mercurians, for Mercury lies too near the sun and is too small to possess an atmosphere. The same goes for Venusians, to a very large extent. Venus is too hot and the atmosphere is highly corrosive. Jovians and Saturnians are also out since both planets are high-gravity gas giants. This effectively eliminates Uranians and Neptunians as well. Last, and by no means least, there is Pluto, outermost planet of the Solar System. Plutonians, if they existed, would be miserable creatures, doomed to exist forever on a dark, airless, frozen world on the very shores of the great interstellar ocean. What we have said of these solar planets applies equally to similarly placed planets in other solar systems. Let us therefore be thoroughly honest with ourselves. No life is ever likely to appear at such spots, or if by some queer freak of fate it does, it is not going to survive for very long. From such planets no indigenous starships are going to set out across the great divide. This last remark applies with equal force to watery planets and any fishlike beings thereon. It might be possible for fish to develop a greater degree of intelligence than that possessed by the terrestrial varities, but this is unlikely to lead to a great icthyological civilization. Advanced societies require minerals and metals. Just how does one make fire, let alone smelt metal, under the ocean?

Life capable of doing the things we have been considering in the earlier chapters of this book must originate on terrestrial-type planets. This is not to suggest that these remote worlds are going to be carbon copies, exact replicas, of Earth. There will undoubtedly be differences and some of these differences will be quite marked. In their general essentials, however, they will be terrestrial, i.e., small, but not too small; have moderate gravity, considerable land surface, and a breathable diluted oxygen atmosphere; and lie neither too near nor too far from their central suns.

The stars in our galaxy are grouped in two assemblages known to astronomers as Population I and Population II respectively. Those in the former category (which includes our Sun and most of the stars in the spiral arms of the galaxy) were formed from gases that contained the heavier ejector of older Population II stars. The presence of this material is clearly indicated in their metal-rich spectra. In fact, the heavier elements are more abundant by a factor of between 100 and 200, which is very considerable. The consequences of this much greater preponderance of heavy materials are many and varied, but probably the most important is a tendency for the protostars or gas clouds to break up into smaller units, i.e., an enhanced tendency for them to condense into multiple stars and planetary systems. So far as double and multiple stars are concerned, this is well substantiated by observation. It is reasonable therefore to anticipate the occurrence of planetary systems (and hence of life) among Population I stars. Since Population I stars are found within the spiral arms of the galaxy and those of Population II toward, and in, the center hub, it is in the spiral arms that we can most confidently expect to find life and other civilizations. We will therefore eliminate Population II stars and the central regions of the galaxy from our considerations and concentrate instead on Population I stars and the spiral arms. (Interested readers should refer to Appendix I for further details concerning this aspect.)

Stars are grouped by astronomers and astrophysicists into spectral classes, these being designated by specific letters. The original intention was that these letters should follow the normal alphabetic sequence, A,B,C,D,E, etc. For a variety of physical reasons, all sound, it became necessary after a time to shuffle them around so that eventually the complete series ran: W,O,B,A,F,G,K,M,R,N,S (the remainder being eliminated.) This is admittedly an odd sequence to remember but astronomers, being after all quite human, were quick to devise a mnemonic that soon became immortal. *W*ow *O*h *B*e *A* *F*ine *G*irl, *K*iss *M*e *R*ight *N*ow *S*weetest.

The sequence has never since been confused or forgotten—nor is it ever likely to be!

Each spectral class from B to M inclusive has been further divided into nine subunits, e.g., B0–B9, A0–A9, F0–F9, again for distinct and definite physical reasons.

Very large hot Population I stars, i.e., those of spectral classes B0 to A5 have a relatively short life on the stellar time scale (10 million to 2,000 million years) before becoming unstable. Such stars offer insufficient time for life to develop. From about F5 to K5 the period of stellar stability increases from 6,000 million years to around 100,000 million years. It is among planets of these stars that life has the best chance of development. As if to confirm this we find near the middle of the group our own sun, classification G0.

Stable existence of Class F stars (F0–F5) ranges from 4,000 million to 6,000 million years. This is regarded as a minimum threshold with respect to the time needed for the development of intelligent life. The civilizations in F0–F5 star system must therefore relatively soon face the prospect of their sun's instability as it slides from premature middle age to premature senility. This involves swelling up into a red giant star, a bloated thing of enormous proportions, which consumes its inner planets and roasts its outer ones. Such unfortunately placed civilizations must therefore perish or find sanctuary in another star system. Here then we see the need for haste in their development of interstellar travel and the extreme motivation for taking over another people's world—by force if necessary.

The (relatively) cool, reddish stars of spectral classes K and M have, on the other hand, very long stable periods. Around such stars life and intelligence have virtually unlimited time to develop and flourish. Because such stars are cooler than our Sun, any life-bearing planets will have to lie closer to their respective parent suns. Planets of K and M stars at distances similar to Earth's and Venus's from the Sun (93 and 66 million miles, respectively) would be as cold as Mars and the asteroids.

As we move down the scale of spectral classes from G (actually from about G4) life-bearing planets will tend to have a progressively colder climate. Because of this, life processes are retarded, and so it takes progressively longer for higher life forms and civilizations to develop. Balancing this out is the fact that the greater time requirements are easily met because of the longer stable lifetimes of such stars. Thus around G5, K, and perhaps even M stars there could exist life forms and/or intelligences much older than any here on Earth, assuming of course that the stars concerned are older than the Sun. This does not necessarily mean they would be more advanced. If the star is of an age comparable with that of the Sun, life may have existed on one or more of its planets for about 3,000 million years, as on Earth. Because of slower development it could be much *less* advanced.

Another factor of considerable importance is the plurality of inhabited planets in a system, i.e., inhabited planets in which civilizations arose *independently*. Plurality is considered a more probable state of affairs as we move up the scale of spectral classes from G2 to G0 and into F, but much less probable in classes G4 to K and onwards. Current reasoning here is that two or more civilizations in the one system would lead to an outlook

markedly different from our own, notably with respect to social
and philosophical attitudes. It would also accelerate considerably the evolution of simple interplanetary flight capability.
Fusion of these attitudes with such capability would, it is supposed, lead much more rapidly to a state of space-minded
enlightenment and psychological preparation for eventual interstellar operations. Alternatively, the existence of at least one
other habitable planet in a system having only a *single* civilization would bring about similiar attitudes. Only in some respects
are these premises acceptable.

In these pages the theme has been that of possible interplanetary war. So far as we on Earth are concerned, this must, of
necessity, mean interstellar war, since there are no other races in
the family of the Sun to do battle with us. Had there really been
a more advanced race on Mars, it is there that the most immediate threat to terrestrial civilization would have appeared. It was
just this very fact that led H. G. Wells to write his *War of the
Worlds* in 1897. To a highly technological civilization on a dying
Mars, Earth, with its vast oceans, its prolific mists and clouds, its
rolling rivers and fertile land masses would have seemed idyllic.
It is interesting to reflect that if such a race had existed, Wells'
book might never have been written—fact could have preceded
fantasy and Martians might have ruled Earth today.

Equally interesting are reflections on what might have trans-
pired had Venus been a pleasant, uninhabited planet instead of
the uninhabitable furnace it really is. With a terrestrial civilization growing on Venus, trade and traffic between the two sister
worlds of the Sun would certainly have accelerated the space-
conditioning process, thus bringing forward the era of interstellar exploration. This assumes, of course, that with the passage of
time the two worlds had not quareled. Since terrestrial nations
have been known to fight one another long and viciously, such
reprehensible behavior could easily have been given an interplanetary dimension—Venus doing a "Pearl Harbor" on Earth
or vice versa. At first Earth would almost certainly have regarded Venus as a colonial world. No doubt politicians on
Earth, running true to form as ever, would have seen in this a
heaven-sent opportunity to tax the new Venusians with increasing vigor so as to finance their own pet schemes or further their

own ambitions. Such a situation can, understandably, soon become very sour and the consequences far-reaching.

Between the extremes of the F and the K class stars lie those of class G. These offer intermediate conditions with respect to stable stellar lifetimes and evolutionary pace. They are not more likely to have planets than are the K class stars, but in class G there is a greater possibility of abundant life on at least one planet in every system. On the planet with most abundant life, intelligence is likely to appear. In G class systems such intelligences have sufficient time to develop into technological civilizations because of the long stable lifetime of the central star.

Though this has been rather a sketchy outline of the position as it appears today, it will nevertheless be apparent that a correlation exists between planets, life, and form of civilization on the one hand, and the spectral type of the star on the other. The correlation is an aspect that could be gone into very profitably and in considerable depth—but, the present pages are hardly the place for an exposition along these lines. Nevertheless, we ought at least to summarize the conclusions reached so far since these are highly relevant to our theme:

1. Taking the galaxy as a whole, centers for the initiation and development of life are much more probable among the spiral arm (Population I) stars. Significantly, our Sun is an average Population I star within one of the spiral arms.
2. The possibility of planetary systems increases significantly beginning with the middle and late class F stars, the rotational velocities of which are comparatively low. Slow rotational velocity is seen as a sign that a star has spawned planets. Were the star unable to shed its excess angular momentum in this manner, its physical integrity might be threatened.
3. This low rotational velocity trend continues throughout the G, K, and M spectral classes.
4. On this basis it is reasonable to expect that single G, K, and M class stars with masses comparable to that of the sun (0.5 to 1.5 solar masses) have planetary systems.
5. Most of the middle and late class M stars within 23 light-years (7 parsecs) of the Sun are quite small ($<$ 0.5 solar

mass). Such stars have a greater abundance of metals and have therefore a relatively high density. The dimensions of the protosolar clouds developing into such stars are less. This means a lower rotational velocity buildup and hence the likelihood of material being cast off to condense into planets. Stars of this class are also cooler and therefore have a smaller ecosphere (region surrounding a star in which conditions are right for life to develop).

6. Intelligent life having manipulative capabilities is most probable around the G and early K class stars. This assumption is based on astrophysical and exobiological reasoning and not on the single fact that our own sun falls within the category. The fact that it does, however, fits in nicely with theory.

7. Late class F stars could conceivably represent the greatest invasion danger to us. This is directly related to the shorter life expectancy of such stars. Among civilizations on planets of these stars there would be a real and growing urgency to develop and perfect interstellar flight.

In the case of stars with longer stable lifetimes (G, K, M), the impetus toward interstellar capability might in the first instance accrue from scientific curiosity and a desire to explore other parts of the galaxy. With late class F stars the name of the game would more likely be survival of the species—the most powerful and potent impetus of all. The urgency must, of course, be seen in its proper context. We are not trying to suggest that beings on a planet of a class F star are saying to one another: "Our sun blows up in ten years so we had better get out of here and find another solar system while the going is good!" The time scale, we must hasten to add, is somewhat longer than this. Whereas stars in the other classes deemed appropriate for life initiation and development have lives that allow from 10,000 to 70,000 million years for this process, late class F stars only permit 4,000 million years before they set out on the suicide path to red giant state, consuming their inner planets and roasting their outer ones. On creation's time scale this is all too brief, and were it not for the fact that class F stars are hotter (and therefore more conducive to life processes), it would be too short. Life on the planets of class F stars must be life in more of a hurry. Such beings then must, on this time scale, find a haven soon. They therefore have to be ruthless if a backward race orbiting a "safe" star stands in their way. There is, however, a built-in safety factor for us in that class F stars are comparatively rare. Moreover, the likelihood of their having planetary systems is considerably less than in the case with G and K type stars.

In the immediate vicinity of our Sun (i.e., within 20 light-years) stars with planets suitable for habitation by intelligent beings are relatively few. They are the following ones:

Star	Spectral Type	Distance (light-years)	Remarks
Alpha Centauri	G0, K3	4.3	Nearest star. Though a binary, the separation between components may permit planets.
Barnard's Star	M5	6.0	Cool, small ecosphere
Lalande 21185	M2	8.2	Cool, small ecosphere
Epsilon Eridani	K2	10.8	Any innermost planets could be habitable.
61 Cygni	K5, K8	11.1	
Procyon	F5	11.3	Hot star. Life could develop very quickly.
Epsilon Indi	K5	11.4	
Tau Ceti	G4	11.8	Solar type. Very favorable.
70 Ophiuchi	K1, K5	16.4	Similar to Alpha Centauri system.
Eta Cassiopeiae	F9, K6	18.0	Strong possibility. Adequate binary component separation.
Sigma Draconis	G9	18.2	Strong possibility. Single solar type star.
36 Ophiuchi	K2, K1, K6	18.2	Triple star system. Could have planets. Stellar components well separated.
Delta Pavonis	G7	19.2	Strong possibility. Solar type star.

Here then are thirteen stars, all relatively close to us, that could harbor races of high intellectual and technological capacity. In this context (and inevitably) life is more probable on some than on others. Single stars are always a safer bet. Planets oribiting double or triple star systems could have very odd orbits that might easily render them alternately too hot and too cold for either life initiation or its subsequent development. If, however, the stellar components have sufficiently wide separation, planets *could* orbit individual components thereby obviating this difficulty. It will be noted that the list contains only one F class star. This is Procyon, principal star in the winter constellation of Canis Minor and one of the brightest and best known stars in the northern hemisphere.

"Creeping colonization," a feature suggested earlier, could most easily manifest itself from one or more of these thirteen stars on acount of their "close" proximity to our sun. This might involve the very early stages of such colonization beginning from one of these stars, or merely a further phase of an ongoing process that had its origins in a more remote star system centuries or millennia before.

The second star on our list is Barnard's Star, only six light-years distant. It is a relatively cool star and consequently its ecosphere or region appropriate to life initiation and development is small. Being a single star it is nevertheless more likely than the nearer Alpha Centauri system to have planets. It could therefore very easily be the nearest other solar system to our own and therefore also house the nearest other civilization to our own. In the light of these facts it might be a good idea to take a somewhat closer look at this particular star.

It was as recently as the summer of 1963 that Dr. Peter Van de Kamp, director of the Sproul Observatory in Pennsylvania, announced the discovery of a small planetary companion to Barnard's Star. Since then Van de Kamp has announced, after further observations, the presence of yet another unseen planetary companion. The star itself is invisible to the naked eye, being of magnitude 13. It is a small, red dwarf with a surface temperature of 2900°K and represents one of the commonest types of star in the entire universe. Eddington later estimated its diameter as approximately 0.15 million miles and its mass as 15 percent that of the Sun. At that time its only claim to distinction was the fact that it had the greatest proper motion of any known star (10.25 seconds of arc per year). Understandably it was soon nicknamed "The Greyhound of the Skies" by astronomers and space buffs.

What Van de Kamp actually announced in 1963 was the existence of a minute 24-year cyclic perturbation in the proper motion of the star. Such a movement could only be due to the presence of an unseen orbiting companion, i.e., a planet. This planet was estimated by Van de Kamp as having a mass 1.5 times that of Jupiter (i.e., 0.15 percent solar mass). It was, he claimed, moving in an elliptical orbit having a mean distance from the star of 400 million miles. Previous work along the same lines had indicated the presence of "companions" to the stars 61 Cygni, 70 Ophiuchi, Ross 614, and Lalande 21185 (three of these are on our list of thirteen stars). In all these cases, however, the "companions" are more likely to be 'substars' than mere superplanets. In fact, so far as Ross 614 is concerned, the companion radiated sufficiently to appear as a faint photographic image.

In June 1969, Van de Kamp and his colleagues claimed that the motion of Barnard's Star (or more correctly the irregularities in that star's proper motion) could best be explained by the presence of two planets, which they designated as B1 and B2, both orbiting in approximately the same plane (just as the planets in our own Solar System do). B1 was estimated as having a mass 1.1 times that of Jupiter, a near circular orbit of approximately 400 million miles and an orbital period of 26 terrestrial years. B2, on the other hand, appears to have a mass 0.8 times that of Jupiter and also describes a near circular orbit. The orbital period in this case is put at 12 terrestrial years.

It is fairly safe to say that B1 and B2 represent true planets and are probably the *major* planets of a solar system containing several smaller members also. If this is so, then there is a reasonably close resemblance to our own system, which comprises four major planets (Jupiter, Saturn, Uranus, and Neptune) and five smaller members, (Mercury, Venus, Earth, Mars, and Pluto). Unfortunately, the degree of perturbation caused by such small fry on the proper motion of a central star is very slight and beyond our ability to detect at the present time. Astronomers of more advanced civilizations may not suffer this handicap and some could be well aware of the five smaller planets that also orbit that star we call "the Sun."

In theories relating to the formation of solar systems it is generally accepted that composition and mass disposition of the small and the large planets are not due to chance. They almost certainly accrue from the condensation of matter remaining after the condensation of the central star. It is believed that metallic and rocky matter condenses first, with liquids (e.g., ammonia, water) condensing later. The former, of course, produce the small planets (in our case Mercury, Venus, Earth, Mars, and Pluto), and the latter the large "gas giants." It may be, therefore, that similar processes were at work in the case of Barnard's Star, resulting in the giant planets B1 and B2 and a number of smaller rocky planets of a more terrestrial nature orbiting closer to the central luminary. (Readers may be quick to point out that, although in the Solar System, Mercury, Venus, Earth, and Mars orbit close to the sun, small, (presumably) rocky Pluto does not. The point is valid, illustrating once again that Pluto is something of a hybrid and a mystery— probably a moon of Neptune, which somehow contrived to break loose and live a planetary life of its own.) Lyttleton has postulated that Earth, moon, and Mars formed simultaneously as a consequence of a larger body dividing itself like a cosmic amoeba into three. Mercury and Venus could conceivably have resulted in a similar fashion. If this theory has validity, then it is perfectly feasible that Barnard's star could have two, three, or more small inner planets of a roughly terrestrial nature.

It is on such inner planets that life, if it is present at all, must exist. In the case of Barnard's Star these planets will orbit much closer to the star than do the inner planets of the Solar System with respect to the Sun. This is a perfectly natural consequence of a smaller star and one emitting only 0.001 of the radiation coming from the Sun. The fact that planet B2 is only about 250 million miles from the Star and yet is presumably an "ice giant" lends emphasis to this fact. Because of the very low radiation, the ecosphere must inevitably be very small. In the Solar System the ecosphere begins just beyond the orbit of Venus and reaches almost to the orbit of Mars. Only in that region are conditions sufficiently temperate for life. The outer limit of the ecosphere in the case of Barnard's Star is probably less than the orbit of Mercury. For a planet of that star to receive heat and radiation equal to that which Earth receives from the Sun, it would have to lie only three million miles from the Star!

Such a planet is possible, but chances are that its close proximity to the star would result in its having one hemisphere eternally turned toward it and the other away from it. There would thus be one hemisphere of warmth and unending day and another of unbelievable cold and eternal night.

One other snag toward life and biological processes on such a world should be considered. Stars of this type are prone to flaring. Our own Sun from time to time gives rise to solar flares. On a world such as Earth, 93 million miles distant, these are more or less harmless. On Mercury, only about 30 million miles distant, they would be distinctly lethal to any life. To a planet existing only a tenth of that distance from its star, flaring would be utterly inimical to life and would in fact inhibit the life initiating processes going on in any primeval "soup."

In our list of thirteen stars then, Barnard's Star must come very close to, if not actually at, the bottom. Nevertheless, it is close to us in the interstellar sense, and what we have said regarding its planets, their sizes, relative dimensions, and orbital positions may well apply to all the single stars in that list and perhaps to the multiple systems as well assuming the stellar components lie sufficiently far apart. We must never forget also that life in the universe might be that bit more adaptable than we think. A race on a planet more than three million miles from Barnard's Star might be able to exist at temperatures below those that we believe possible and at the same time be able also to withstand, perhaps even absorb beneficially, intense bursts of radiation. In our list, however, Epsilon Indi, Tau Ceti, Eta Cassiopeiae, Sigma Draconis, and Delta Pavonis are the most likely quarters for the rise of alien civilizations, and Procyon seems the one from which a civilization might be forced to migrate—perhaps with disastrous results to others.

At this point we would again remind readers that colonization and conquest by intelligent beings able to "short-circuit" space by hyper-dimensional techniques is not going to be inhibited by distance. In such circumstances the foregoing list is irrelevant, for the threat could arise anywhere within the galactic arms.

About one point we can be quite certain. The Sun is not the only star in the galaxy to have planets; Earth is not the only planet in the galaxy to harbor intelligent life. It is only in recent times, i.e., in the past few decades, that these concepts have found acceptance. It is interesting to reflect how through recorded history cosmological beliefs have changed despite the forces of reaction and prejudice. First it was firmly held that the Earth was flat. The belief that it was round was not allowed to develop overnight. That Earth was not the center and prime body of the Solar System was another fact a long time in passing (the church of that day carries much responsibility for that). Next it was shown that our Sun is just a very average star, one of a hundred thousand million constituting the Milky Way and, far from being at its center, lies two thirds of the way out towards the perimeter. Then in 1914, it was shown categorically by Edwin Hubble that several of the faint diffuse bodies apparently among the stars of the Milky Way were, in fact, galaxies in their own right, many being larger than our own and all lying millions of light-years from the Milky Way. The next advance in thinking is the existence of other solar systems and, consequent upon that, the concept of other civilizations, many much more advanced than our own. This in turn leads to thoughts of interstellar travel.

And thus, just as we will one day explore the galaxy, so today some will already be doing exactly that—in many instances for purposes of peaceful exploration and the extension of knowledge, in others to escape from a dying sun or a dying planet.

19. MOTIVES FOR SPACE WAR

Having looked closely and dispassionately at some possible aspects of interplanetary war we should now perhaps give a little thought to the reasons for such a calamity—for calamity it most assuredly would be.

Wars do not just happen. Something causes them. In the not too distant past it used to be said that war was merely an extension of diplomacy. This was something of a facile view, perhaps even a false one, since war generally follows upon the failure of diplomacy. In the case of a star war, peoples (probably of different form) battle with one another either in space, on the surface of planets, or both. If they have had no previous contacts or knowledge of one another, no inkling of each other's language, then questions of diplomacy do not arise. This aspect can only come to the forefront if the war is between peoples of a civilized kind who have traded representatives and generally conducted some form of negotiating procedures. Such thoughts bring us very close to the galactic federations so beloved of many science fiction writers. This is not to suggest that galactic federations and the like do not exist. Perhaps they do, and for all we know at this very moment one or two of them in far-flung corners of the Milky Way or deep within the star mists of its center are battling viciously to the death. It is a less than pleasant thought, but as we mentioned earlier we may have to accept the universe as it *is*, not as we would *wish* it to be. On dark frosty nights when the heavens are powdered with stars it is intriguing to gaze up into the seemingly tranquil depths and contemplate something of the odd dramas perhaps being enacted there. The strange, unreal battling starships that we have just left on the television screen could be very real and perhaps of even more bizarre form.

Aggression and invasion have been unhappy facets of terrestrial life since the dawn of history and probably a lot earlier. We on Earth then are in a position to speak freely and with considerable authority on the reasons for these things. During very early times warfare was largely a feature of tribalism (there are those who claim this is still so). The main purpose of tribal war was simply that of plunder, involving goods, territory, livestock, and slaves. As we grew more civilized (supposedly), other reasons for war and conflict soon began to appear—ambition, lust for power and dominion, aggrandizement on a vast scale, revenge, and the desire to reverse the decisions of earlier contests. In more recent times a fresh and even more obnoxious parameter has crept in and is now rearing its ugly head—that of political ideology, generally a blind and near fanatical adherence to a certain cult or "ism" of the extreme left or right, most often the former, masquerading as "democracy!"

Today an entirely new factor has arisen. More correctly, it arose a considerable time ago, but the world and its leaders were blissfully unaware of the fact. Now it has come home to roost. It will not go away—not, that is, unless positive steps are taken. Soon all but the ignorant will see it—and admit to seeing it. It is the one perhaps most likely to result in widespread, devastating, internecine war among the peoples of this planet—overpopulation, gross and escalating overpopulation. Inevitably in its wake come the associated ills—overcrowding, famine, pollution of wind and wave, increasingly heavy denudation of limited, nonrenewable resources.

If other galactic civilizations have the means of interstellar travel (and a steadily growing number of scientists believe this to be likely) it is chiefly for this reason they would go to the trouble of crossing the immense cosmic deeps with all its attendant difficulties and dangers. Despite a cult of science fiction writers long wedded to the idea of star wars fought for purposes

of galactic dominion or ideological ends, an invasion of our
world by a race from one of the stars is much more likely to
occur because the planet (or planets) of the home system has
become grossly overcrowded and there are no suitable other
planets within that system to which the excess population can
migrate. If at that point in their history they happen to have
acquired, in one form or another, a method or technique by
which they can cross the interstellar abyss to the fresher planets
of a younger sun then, quite clearly there is but one logical
course open to them. That course expressed in its simplest and
starkest terms is the large-scale transference of an appreciable
portion of their population to such planets. Such an issue, sooner
or later, becomes one of life or death. Morals, ethics, issues of
right and wrong play little or no part in it. And, let us be honest,
terrestrial man would act and think in precisely the same manner if ever he were to be confronted by like circumstances! Other
reasons for an invasion from the stars do undoubtedly exist. To
the writer they appear to possess a considerably lower priority.
He could, of course, be wrong. Nevertheless the population
factor seems an excellent and highly valid one with which to
start.

    At present we are hearing more and more on the subject of
our world's exploding population and its steadily diminishing
resources. Those who draw attention to the dangers are as often
as not, labelled "doomsday merchants;" those who dismiss it all
as panic-mongering, "ostriches with their heads in the sand."
This leaves a mass in the middle not really knowing what to
think but believing that it wouldn't make a great deal of difference anyhow. What then is the true position and what are the
dangers—real as well as potential? Though our theme in these
pages is primarily that of civilizations on other planets and how
these might one day impinge on our own, we must always
remember that ours is just one of the planets in the galaxy and
what is happening, or is about to happen here, could easily be
happening, or already have happened on others. It therefore
makes sense to review and analyze the situation as it applies here
and, if possible, to predict the logical outcome. Once we comprehend these things and their import, we may better be able to

appreciate how compelling could be the forces leading the peoples of one planetary system to attack those of another.

According to historians, the population of the Earth during Roman times was about 150 million with a yearly population increase of 0.07 percent (105,000 per year). The present day population of our planet is 3,650 million, over 24 times as great and the yearly increase rate 2 percent (78 *million* per *year*). In other words, Earth is now gaining population to such an extent that nowadays only two years are required to add to the total a figure representing the Earth's entire population during Roman times. If that is not a sobering reflection it certainly ought to be!

Various gimmicks can be carried out involving statistics such as these. A recent one took the Earth's present population figure (3,650 million) and, on the assumption that the average human weight is 100 pounds (this allows for children and infants), put the total mass of human bodies at about 180 million tons. If the present trend continues, then in approximately 35 years the population of our planet will have doubled (7,300 million) with an aggregate flesh, bone, and blood mass of 360 million tons! The figures for the present are statistically accurate; those relating to 35 years ahead are a direct extrapolation. These are not contrived doomsday figures. They are derived from cold, legitimate statistics. Popular science writer Isaac Asimov has recently extrapolated this trend to the limit. The result is frightening. At the present rate of increase in human population the mass of humanity will, by 6826 A.D., that is in 4,847 years, *equal the mass of the known universe.* For the record, the universe consists (very roughly) of a hundred billion galaxies each of which contain on average a hundred billion stars about the size of our Sun. Since the mass of the Sun is in the region of 2.2 billion, billion, billion tons, the mass of the entire universe must be about 3×10^{50} tons (i.e., 3 million followed by 44 zeroes!).

This, of course, is only a mathematical exercise. Clearly, other physical factors must eventually intrude. An extrapolation such as this is rather akin to that concerning the victim of a tropical fever. If his temperature continues to rise in the way it has been, it is only a matter of time before he ignites spontaneously! The value of Asimov's extrapolation lies in the fact that it points most eloquently to the existence of a highly dangerous trend. While we worry about the nuclear bomb and the awful Pandora's box that it has opened, an even more dangerous bomb is ticking away slowly—and even more surely.

Let us return again for a few moments to that fantastic extrapolation. Many might be inclined to say: "So what. Why worry? The year 6826 A.D. is a long time away, even if a thing like that *could* happen." On the human life scale it is indeed a long time away. It represents an era that only our distant descendants will see. If we regard 25 years as representing a generation, then a period of 4,856 years represents 194 generations— our grandsons and granddaughters with 191 "greats" in front of the relevant term. But, and this is the important point, descendants, actual descendants of ours *could* see it. It is therefore not an unimaginable period of time and this fact should serve to indicate the highly dangerous trend already well in existence. This trend, unless checked, could, long before 4,856 years have elapsed, render our planet ungovernable, unmanageable, and finally uninhabitable!

If we happen to be of a mathematical mind and choose to delve more deeply into the figures, we find that in 1,560 years, i.e., by 3,530 A.D., the mass of humanity should *equal the mass of the Earth*. This obviously sets a limit to the process though hardly a very realistic one. Clearly, the process must stop for one reason or another long before this. Let us therefore look at the biological implications for a lead.

The mass of living tissue presently upon Earth cannot continue to increase for the simple reason that only a proportion of the basic energy source, sunlight, can be utilized in the process known as photosynthesis. This severely restricts the amount of new living plant tissue that can result in any single year. Animal life depends ultimately on plant life (carnivores eat herbivores, which eat vegetation) and consequently the former cannot expand indefinitely either. Are then the prophets really "doomsday merchants" after all? Will nature eventually balance things out and correct our mistakes for us? Indeed she will, but hardly in the way we might be tempted to suppose. By 2436 A.D. the population of Earth at present trends would be 40 trillion (8,000 times the present figure). This works out to 200,000 beings per square mile if we include *all* portions of the Earth's land surface. By then the last plant would have been consumed by the last herbivore. Soon after, in relative terms, the last herbivore would be consumed by the last carnivore. After that nothing—except perhaps cannibalism! Before that impasse we would most assuredly have been compelled to take certain positive and exceedingly drastic steps.

The only really feasible course of action would be effective birth control on a large scale. So far we have not accomplished a notable degree of success in this field. Perhaps in time increasing realization of the danger will help. We can only hope. The only vaild alternative is more "Lebensraum". Eight other planets orbit the Solar System. Unfortunately, with the exception of Mars (a highly doubtful exception) these are stark, forbidding worlds. That, at best, leaves only the Moon and Mars. To neither quarter can we ever hope to transfer sizable portions of the Earth's excess population. So back to square one—which, in fact, we never really left!

If, however, the attainment and development of interstellar travel potential precedes the disintegration and subsequent demise of our civilization and society because of the starvation and wars (most likely nuclear) that overpopulation would bring, then a lifeline will have been flung to mankind. At the other end of that lifeline is the possibility of planets appropriate to our requirements at an attainable distance. Unfortunately some, perhaps most, of these planets, could contain indigenous societies. This could be rather rough on them. A drowning man clutches very fiercely at any chance of life—sometimes to the extent of drowning another life.

Overpopulation, and the consequences accruing directly from it, would then seem to represent the major cause of an invasion from the stars. This does not mean that we should entirely overlook other possibilities. Alexander, having conquered and brought under his sway most of the then known world, was reputedly saddened because there were no "worlds" left to conquer. Rome grew from a small community on the banks of the River Tiber to become the Eternal City and the center of the greatest empire which, up until then, the world had ever seen. Its rulers did not actually, in physical terms, *require* more and more territory, a fact that did not inhibit their acquiring it—and by force of arms at that. They may not have required more territory, but it brought great rewards to Rome both in material riches and in enhanced prestige. The greater that the might, power, and dominion of Rome grew, the more rewarding and satisfying it was to be a citizen of Rome. Down the long road of history the process continued. When the Romans quit Britain the land was invaded in turn by Angles and Saxons from northern and central Europe. Jump several centuries and we see the notable attempt at dominion by Napoleon Bonaparte. Another century and a half brings us to one of the greatest megalomaniacs of all time, Adolf Hitler, whose thinly disguised attempt to rule the world came uncomfortably close to success. Since the close of that sanguinary and destructive chapter, ambitious characters waving the red flag of communism have taken the stage. Their motives would not seem to be greatly dissimilar, despite loud protestations to the contrary both from them and their puppets in other lands. By and large, human nature, so far as terrestrial man is concerned, doesn't ever really seem to change much. On the whole, however, the human race does seem to be advancing spiritually and morally, though the advance cannot be compared with that made by technology. Here we appear to have a classic illustration of the tortoise and the hare. In the fable the tortoise actually won. This seems highly improbable in the case of man. It appears at times that there will always be one man or a group of men hell-bent on achieving power for the mere sake of it. And by skillful manipulation of mass thought they can generally secure sufficient followers and disciples to have a go at it.

The question that inevitably follows is whether or not some, perhaps even all, alien races have this same inherent weakness. It is a question that for the foreseeable future we cannot hope to answer, but it does seem reasonable to assume that at some points within our galaxy (and in others) there will be men or beings who lust for power. If such beings are members of races without, and nowhere near, the attainment of interstellar potential, then obviously we have nothing to fear. If races so afflicted have achieved and developed such potential, we must sincerely hope that they all exist on the other side of the galaxy, though highly sophisticated techniques of extradimensional transit might well invalidate the security that distance would normally be expected to bring.

It is interesting to reflect on a hypothetical instance of cosmic aggression as it concerns our own planet. It is necessary in the first instance to make two assumptions—first that Adolf Hitler did, in 1940–41, succeed in extending his hegemony over the entire globe and second, that interstellar travel had, by then, been perfected. Now here we have an interesting situation. Would Hitler have been content to rule just one planet, a mere terrestrial empire, when he could also rule a cosmic empire? He almost certainly would not. And would he have been caring and considerate to the indigenous inhabitants of these other planets? We only need to glance at the record of what he and his cohorts did to the peoples of occupied Europe during World War II to find the answer. But perhaps we should not overvilify Adolf Hitler. Attila, Caesar, Genghis Kahn, or Stalin in the same position of power would almost certainly have acted in like manner. Mercifully for any neighboring galactic peoples the gift of star travel was not bestowed on these men or others of the same breed. We must hope, very sincerely, that when at last the gift is bestowed on terrestrial man, as one day it almost certainly will be, he will not use it to enslave or destroy less advanced peoples he finds in the cosmos. At this point the reader will probably recall that we envisaged doing something along these very lines should we at some time in the future find it necessary to move surplus population in a future era or migrate "en masse" because Earth was dying. There is, however, something of a difference. In the case we have just been discussing, the thing is done merely to satisfy a craving for dominion and a love of power; in the other it is directly attributable to desperation. In neither case is the act morally defensible. Wholesale genocide or enslavement never can be. In the latter instance, though, there is at least some rationale for the act.

We should perhaps be careful that, by accepting the possibility of galactic empires, we do not lean too far toward the world of "pulp" science fiction where the most incredible looking starships slice viciously and searingly at one another with equally incredible colored rays, where physically perfect terrestrial men do battle with strange alien beings, and where equally perfect young terrestrial women (often in rather revealing and impractical raiment) await rescue from the alien hordes. In this sphere of things our galaxy is almost invariably portrayed as full of cosmic dynasties with one or the other trying to extend its dominion throughout the entire Milky Way. When we consider that our galaxy comprises 100,000 million stars and is 100,000 light-years across, this does seem a rather improbable state of affairs. We can, with much more justification, envisage a form of cosmic dominion embracing a "close" group of stars. It is difficult to foresee any civilization extending its hegemony over *all* the suitable planets in the galaxy, though a policy of gradual colonization could lead to ever widening bounds. In time the influence of two or more groupings could impinge upon one another. Then indeed a form of star war, a contest between two galactic empires, could ensue. It would certainly not be our plan, and even less our desire, to blunder inadvertently into anything like that!

In his *Relations with Alien Intelligences,* Dr. Ernst Fasan outlines eleven "Laws." In effect these constitute a sort of cosmic international law. They are eminently reasonable but, as with all laws and international agreements, would require more than unilateral observance. In some instances one such law tends toward incompatibility with another as for example Rule 7 ("Every race has the right to defend itself against any harmful act performed by the other") runs counter to Rule 8 ("The principle of preserving one race has priority over the development of another race"). By Rule 8 future man, from an exhausted and overpopulated Earth, has the right to take the world from a still developing race—and by Rule 7 has no right to complain if he gets a bloody nose trying to do it!

Rules 5 and 6 reinforce Rule 7 and the defenders:

Rule 5. "Any act which causes harm to another race must be avoided."

Rule 6. "Every race is entitled to its own living space."

Interplanetary aggression is also opposed by Rule 3, which states: "All intelligent races of the universe have, in principle, equal rights and values."

Unfortunately, as most of us are only too well aware, formulating a framework of laws is relatively easy; securing its acceptance by all parties much less straightforward. We could also add that ensuring its observance by those who have already accepted it is no easy task either. What right have we to assume that galactic law would differ in these respects from international law? The position is further complicated by the possibility of strangers from a remote part of the galaxy who are not aware of the law—and who probably would not wish to be.

It is sometimes said that aliens would not attack us, having advanced so much further along the road from barbarism to advanced culture. This is a highly risky philosophy at best. During the present century a number of terrestrial nations have attacked weaker neighbors. In those instances the aggressors were supposedly cultured nations. There are, we must suppose, different kinds of barbarism. There is the basic type seen, for example, centuries ago when the Goths and Vandals sacked Rome. But was a Rome that gloried in murderous orgies in the Colisseum really any less barbaric? Rome's barbarism merely flourished amid culture—or vice versa. The barbarism of Nazi Germany existed in a seemingly cultured, civilized state. In this instance we see very clearly what could well be described as the "new barbarism." It is rampant today in a number of countries and with the passage of time seems to be on the increase. It is this other form, this "new barbarism," that could so very easily be part of the fabric of a highly technological civilization reaching us from the stars.

So far as terrestrial society is concerned, it could be said there are members of it who genuinely espouse true culture, those who pay a certain lip service to it, although they are not averse to committing wanton acts if circumstances seem to warrant them, and finally, a brutish component to whom violence and a total disregard for law and order is both a natural reaction and a way of life. This could well be the type of setup that has developed among other cosmic societies. Some may indeed have become utopian in ideal and outlook. From such peoples we would have little to fear—so long at least as a serious threat to their world did not arise. Perhaps in time terrestrial civilization will achieve its utopia, too. At present it appears to be permeated by men of violence who have not the slightest hesitation in murdering innocent people in pursuance of political ends. We have a long way to travel on the road to utopia. In the circumstances could we justifiably complain if an extraterrestrial race descended upon us from the skies and began taking our world by force? We would certainly have to defend ourselves and our world. But the moral issue of right and wrong might be that little less sharply defined.

20. THE BEGINNING— OR THE END?

In this, the final chapter of a soliloquy full of the direst possibilities for our world and our society, it might prove a sound idea to stand back a little, as it were, and endeavor to view the subject more objectively.

Many may feel (and this is fully understandable) that the possibilities discussed at some length in these pages are just too terrible and too fantastic ever to happen. Undoubtedly they represent a most appalling prospect. But is that prospect any more terrible than full-scale nuclear war waged among the nations of Earth? When, in that now curiously remote year of 1906, the late H. G. Wells envisaged atomic bombs dropped by aircraft, many must have dismissed the idea as fantastic beyond belief and been highly relieved they could do so. After the events of August 1945, such comforting and cozy thoughts had to be abandoned. We had instead to live with the reality—and it has not been easy. Yesterday's nightmare had become tomorrow's vision. We listen and seem at times to hear the riders of the Apocalypse heading our way. But the form and the manner of their coming may be rather different from what we have come to dread—which is precisely what this book is all about.

Inevitably, there must be those who, though they concur with the idea of superior galactic civilizations and the feasibility of interstellar travel, believe nevertheless that the galaxy is so vast that our little corner of it must remain forever sacrosanct, that somehow the odds are still very much in our favor throughout the foreseeable future. They could be right of course. They could equally well be most horribly and tragically wrong. The present writer freely and gladly admits that this is an occasion when he would greatly prefer to be proved wrong. To be wrong and alive is, after all, infinitely preferable to being right—and dead!

The basic question remains the same. Would the arrival of advanced aliens from the stars necessarily mean war, invasion, and destruction of our civilization, our freedom—and probably us? Only, as we have stressed, if the aliens (a) had a desperate need for our planet in order to ensure their own survival and continuation as a species or, (b) happened to be galactic empire-builders requiring subject peoples and subject worlds to enhance their well-being and further their grandiose ambitions. A third possibility, that they might desire us to join their galactic federation for mutual benefit, would not necessarily involve us in the horrors of an alien invasion. Of course "persuasion" comes in different guises! The tremendous possibilities for us in accepting integration in a great and benevolent new order would probably be sufficient. Joining a galactic federation could well be infinitely preferable to an Earth ruled by an indigenous tyranny, a possibility that, with every passing year, seems more likely. Being equal members (eventually) of a great, galactic federation has long been a favorite theme of the science fiction writer—and not without reason. Resources at our disposal would be those of a plurality of worlds backed by the skills, resourcefulness, and guidance of peoples our seniors and superiors in so many diverse ways.

We are only able to speculate on the future of our world and its civilization. This is as true now as it has always been. To foretell the future is a power that has not been given to mortal man—and for that we should probably be thankful. Our little planet, third world of an undistinguished star, could remain inviolate for all eternity, it could eventually be integrated into a great and free society of kindred worlds and peoples, or it could be attacked and invaded by a cosmic race made ruthless by reasons of necessity or ambition. To the writer any of these propositions seems valid. None is certain to happen—and none is certain not to!

It is interesting in these closing paragraphs to dwell on the thoughts and words of others concerning the possibilities. Extremely intriguing are the views and comments of Professor Sir Fred Hoyle, one of Britain's foremost astronomers and a man of considerable vision. His belief is that one day aliens *could* move against us but in a manner infinitely more subtle than anything envisaged so far in these pages. What, in fact, he suggests is the possibility of an alien civilization feeding us electronically with information that, if used, could lead us into bringing about our own extinction. It is at once a most profound, novel, and interesting theory. "The distances involved are so vast," he says, "that a physical invasion would be impractical. They would destroy us by means of information beamed our way by radio and picked up by research scientists." Such information, he states, would be as revolutionary to us as present-day man communicating to the ancient Romans the secret of nuclear fission. "We might not be able to resist the temptation," he adds, "or fail to see its inherent dangers—and that could be our undoing!" In chapter 7 we envisaged alien attackers to whom Earth's indigenous population could represent both an embarrassment and a nuisance. Skillfully manipulating our thoughts and actions so that eventually we removed ourselves from the scene would be a convenient and cheap method of achieving that very end. A lifeless Earth would then be there for the taking, when at last the aliens chose to turn their attention to this corner of the galaxy. The difficulties inherent in passing information by radio over vast distances in an intelligble form should not be discounted, the more so if the receiving end (us) were relatively backward. A lack of common language would in itself be an obstacle of considerable proportions. Can we be sure that some cosmic race of massive intellect and technological power could not somehow, electronically, telepathically, or otherwise, "suggest" to us that we do certain things, invent certain items? Suppose the invention and use of nuclear weapons had been "suggested" to us in a way so subtle that we firmly believe the idea to be our own. They then "suggest" that we divide the world into two politically opposed camps. Finally they "suggest" that a preemptive nuclear strike by one side against the other is essential. Fantastic? Perhaps it is, although at times it is difficult not to wonder what the strange urge is in human beings that makes them want to emulate the gadarene swine of biblical fame.

A rather novel view on the subject of interstellar attackers has been expressed recently by Dr. Robert Jastrow of NASA's Goddard Institute for Space Studies. Dr. Jastrow believes that aliens will not resemble terrestrial beings because, as he puts it, "their evolutions and civilizations are so far ahead of our own." He goes on, "They may be black boxes filled with sophisticated electronics or mere 'rivers of electricity' with very little bodily form." Whatever their form, Dr. Jastrow believes that aliens will *not* be hostile. "They will welcome the fresh idea of Earth because our planet is much younger than so many others in the galaxy." The possibility of the ideas of contemporary Earth being welcome to a highly advanced race seems a bit sanguine, to put it mildly. Some of our ideas and actions at the present time are more appropriate to the cosmic refuse heap! Few however, would disagree with Dr. Jastrow when he remarks, "The aliens will be so far in advance of us that their command of natural forces and communications will be far beyond our comprehension."

Dr. Jastrow is also of the opinion that some space beings may already be trying to get in touch with us. "High intensity radio and television signals from Earth have been radiating out into space at the speed of light almost since broadcasting began. These signals in that time have swept past about a hundred stars, some of which may easily be capable of sustaining life. Word has therefore reached these distant beings that there is something going on here."

This possibility immediately invokes recollections of the dire warning sounded a number of years ago by Professor Zdenik Kopal of the University of Manchester, England, one of the world's foremost astronomers. "Should we ever hear the space-phone ringing" he said, "for God's sake let us not answer but rather make ourselves as inconspicuous as possible to avoid attracting attention." Many people (including the present writer) commented at the time that, in a sense, we had already rung *them*. When a civilization discovers and uses radio, it effectively announces its existence to those beings in the galaxy that have the ears to hear and the minds to comprehend. This is particularly true with respect to the ultrahigh frequency channels we have taken to using over the last three or four decades. By now it is perfectly feasible that somewhere, up to thirty or forty light-years out at least, one other civilization either knows or strongly suspects the existence of an inhabited planet in the environs of a bright yellow G class star we call "the Sun." They may not be able to reach us—yet!

Professor Kopal has raised a possibility that we have largely ignored in the preceding pages. We have thought in terms of aliens aiming to take from us by force something that we have and that they require or want—most likely our planet. To a lesser extent we have also envisaged integration, compulsory or otherwise, into a galactic federation or empire, which might or might not be to our advantage. Zdenik Kopal sees the nature of the threat in rather different terms—our being treated as mere laboratory specimens by an advanced galactic society of starfarers. Not unnaturally this leaves us with a highly injured sense of pride—we, the pride of creation as mere laboratory animals—unthinkable! But it is *not* unthinkable. We have to accept the fact that other galactic civilizations could be very much further advanced than our own (this in itself is probably a blow to our pride). Societies spread throughout the galaxy have at least one thing in common with individuals here on Earth—they can have no control over their age. Their respective suns were created at a certain point in time just as we were born in a certain year. Perhaps our feelings of inferiority can be countered by the premise (one just as reasonable) that our galaxy will also have civilizations that are far behind us. Nevertheless, it is a most humbling thought that to certain of our seniors in the universe we could be mere specimens of a low-order life fit only for clinical and laboratory examination. And yet this could be. In an earlier chapter we alluded briefly to a science fiction short story written and published in the late 1930s in which an alien star vessel lowered a huge dragline, scooping up great chunks of our planets geology, flora, and fauna for close examination.

However improbable and impractical such a peculiar method of sampling, here, at least in essence, was that very idea being expressed.

Convinced that sooner or later contact between ourselves and intelligent aliens is "inevitable," Zdenik Kopal warns, "We might find ourselves in their test tubes or other contraptions set up to investigate us, as we investigate insects or guinea pigs." It must be admitted that an alien race so far ahead as to regard us in such a lowly role would certainly be able to cope with us, and sample us, in highly sophisticated and imaginative ways—though it is doubtful if in the circumstances we would appreciate these ways! At first we might not even realize it was happening. Perhaps on the day when ships, aircraft, and their occupants start to disappear without trace as if into another dimension, the era of alien "sampling" will have arrived. It will certainly be time for the alarm bells to begin sounding. (These words should not be interpreted as clandestine support for the "Bermuda Triangle" affair, since such a very big question mark hangs over it. The position would have to be very much more definite and clear-cut.)

It could be argued that, since such processes fall outside the concept of true warfare, they are really beyond the scope of this book. This may well be so. On the other hand, such actions against us could hardly be construed as friendly. The aliens concerned might not be hostile—just utterly and totally indifferent. We scoop up fish from the sea in what can only be described as a thoroughly brutal manner. We then gut and eat them with neither compunction nor compassion. We even pound their bodies into meal for other animals, which we subsequently slaughter and eat. We are not hostile to the fish—but if these same poor fish were capable of rational thought and of understanding the reason for our actions, they would hardly feel that the measures taken against them came under the category of friendly. To them we would quite simply be waging war against fish—which is more or less what we are doing!

Despite all that has been said in these pages, depsite all the seemingly dire forebodings, there is as yet no unequivocal evidence for the existence of aliens. We should dwell on the fact, however, that *absence of evidence should not be construed as evidence of absence!* The position in this respect has been rather well summed up in a recent paper by Michael D. Papagiannis of the University of Boston, entitled "Are We Alone?" To the question "why are they silent?" he replies: "As to why they have not made contact with us, one can think of several answers. The simplest explanation, however, and hence maybe the most probable one, might be that of confusion and indecision. Our hypothetical neighbors were probably acquainted for millions of years with a lethargic Earth inhabited by life forms not worth any effort of communication. Suddenly, in the last fifty years or so, which probably is a very short interval for a well-settled galactic society, they have been confronted with an exponentially mushrooming technological society (aeroplanes, radio-communications, nuclear bombs, space-craft) which undoubtedly must be causing them some serious concern. It is possible, however, that faced with such a sudden technological explosion, a serene cosmic civilization would be perplexed and undecided as to how to handle the situation. They might be debating on whether to crush us or to help us and therefore they might be simply postponing their decision, waiting to see what we are going to do with ourselves."

Part of Michael Papagiannis's paper is concerned with the possibility that the asteroid belt within the Solar System is already serving as "a natural hideout where *they* can remain inconspicuous for a long time until we decide to search for them." This aspect serves as an interesting addition to what we have already said in chapter 17. Moreover, his earlier words provide yet another valid reason for an assault upon us from those outlined in chapter 19.

Easily one of the most pertinent comments on alien life and the risks we might be running came quite recently from none other than Sir Bernard Lovell, director of the famous Jodrell Bank radio-astronomy observatory in England, and one of the world's most celebrated radio-astronomers. His words were quoted earlier in this book, but they are so appropriate that the writer makes no apology whatsoever for quoting them again. It would indeed be hard to find closing words that are more fitting, and the warning we have tried to give more succintly and eloquently expressed. "We must regard life in outer space as a very real and potential danger. You have only to think about the problems of diminishing resources here on Earth to realize that alien civilizations may be combing the galaxy looking for new resources or a new place to settle. They could want something we have got—and they could well have the ability to take it from us whether we liked it or not!"

APPENDIX 1

The theme of this book supposes that alien civilizations are legion throughout both our own galaxy, the Milky Way, and the untold millions of others, seen and unseen. The basic essentials on which this supposition is based are outlined in Appendix 2. Here we would like to set out briefly present cosmological thinking on the origin of the stars around which the worlds containing these hypothetical civilizations abound as well as the reason for the distribution of the particular elements therein. Without the elements oxygen, carbon, and nitrogen, living beings and plants could not have been created. And without such essential minerals as iron, manganese, copper, lead, and tin, only the most primitive and rudimentary civilizations would have been possible—Stone Ages in fact. If all this seems a far cry from the theme of space weapons and space war we would simply remind the reader that without living, thinking beings there could be no space war; and without the essential metals and other elements for weapons, spacecraft, and fuels—it would be equally impossible. Thus if we are to dwell on the theme of possible interplanetary and interstellar wars, we must establish a case, not just for beings in the universe other than ourselves, but for the existence of the materials and facilities to create the necessary implements. Of one point we can be quite certain—a society remains Stone Age irrespective of its intellectual development if it does not have the materials essential to technology. And Stone Age, even Iron or Bronze Age societies, could never leave the surface

of their own planets, let alone fight star wars in interstellar space
or in the environs or surfaces of other worlds.

It is difficult to be very specific regarding the chemistry of the
universe after the explosion of the primordial fireball, i.e., after
the celebrated "big bang." A million or so years on however, the
temperature had dropped to around 3,000°K and matter in the
universe comprised 75 percent hydrogen and 25 percent helium
with small traces of the elements lithium, boron, and beryllium.
Before the conventional, molecular chemistry of today could
come into being, the heavy elements had to be synthesized from
all this primordial hydrogen and helium. This is impossible at
temperatures of the order of 3,000°K. It is almost certain
however, that a transition stage followed, during which the
elements carbon, calcium, iron, and oxygen *were* synthesized
from the original hydrogen and helium. It is interesting to reflect
on how this occurred.

Spectral analysis has shown quite clearly that the stars in our
galaxy may be divided into two distinct groups labeled by astronomers as Population I and Population II respectively. The
latter comprises stars having atmospheres showing the same
abundance of hydrogen and helium as all the stars had when the
universe was about a million-years old. Compared to stars such
as our Sun, these are very deficient in metals. In these regions,
then, we are unlikely in the future to be attacked by indigenous
starships!

Population I stars are a very different proposition, for here the
concentration of heavy elements is about a hundredfold greater.
Our Sun, of course, is a Population I star, as are most of the
others with which we are familiar in the night sky. This distinction, it is thought, can be explained by the fact that the nuclei of
the heavy elements (e.g., metals) were formed deep in the interior of Population II stars. Population I stars were created subsequently out of the material from the violently destructive demise
of Population II stars. The heavy elements once deep in the
hearts of Population I stars were thus loosed into the universe.
This is a fairly satisfactory explanation. We realize that stellar
temperatures are too low for these nuclear transitions to occur in
stellar atmospheres. From this it follows that the chemistry of a
star's atmosphere is the same now as when it was created. The

thought is intriguing, for it implies that every conceivable atom here on Earth at the present time was once an integral part of another star. To sum up, material from Population II stars was used to create Population I stars *and* the assorted debris constituting planets, and the like. Thus, the atoms and molecules of every thing on our planet, *including those of our own bodies* originated not just in the sun but in a star long gone. Perhaps the ancient poet and philospher Lucretius (about 90 B.C.) sensed these facts, however vaguely, when he wrote:

> *This bowl of milk, the pitch on yonder jar*
> *Are strange and far-bound travellers come from far.*
> *This is a snowflake that was once a flame;*
> *This flame was once the fragment of a star.*

Population II stars are the oldest in the universe. Fairly early on in their lives, when gravitational contraction had stopped and the temperature had reached something in excess of 10 million °K, thermonuclear fusion of hydrogen into helium began in the deep interiors of these stars. The hydrogen was consumed very rapidly (in a few hundred million years) until the stellar cores consisted only of helium. These cores contracted under their own weight, raising the temperature to something of the order of 100 million °K. At such temperatures an entire new series of thermonuclear reactions was initiated. The first product was carbon, leading to carbon cores, further contraction and even more massive increases in temperature *(billions* of °K). It was then that the elements oxygen, magnesium, silicon, sulphur, neon, argon, calcium, bismuth, and iron by virtue of thermonuclear fusion between helium and heavier nuclei progressively began. With the creation of iron the stars concerned became increasingly unstable, the heaviest elements of all such as uranium being created during this unstable, final stage.

The cataclysmic events that ensued were either the occurrence of novae or supernovae. These resulted directly from the inability of the stars concerned to resist gravitational attraction any further and their consequent collapse. As a result, a vast amount of gravitational energy was released. This caused the collapsing shell of the imploding star quite literally to bounce back at a velocity of many thousands of miles per second. Implosion had, in fact, resulted in an explosive effect. All that remained of the star proper was a hot and extremely dense core. The gases rushing outward into space became in time the material from which Population I stars (and ourselves!) were created. The residual dense core of the once resplendent Population II star became either a pulsar or a black hole. "Sic transit gloria stella"!

To some extent the creation of Population I stars mirrors that of their Population II predecessors. Primordial or residual gas from Population II stars formed mighty spherical clouds having masses ranging from 0.2 to 10 times that of the Sun. As before, gravitational contraction sets in, and this has the effect of raising the temperature once again to the 10 million °K mark. At this point in the proceedings we once more witness the start of thermonuclear fusion of hydrogen to helium.

There is a distinct divergence, at this juncture, between what happened in the case of Population II stars and now. Without going into the physics of the situation, we can say that from the formation of carbon in Population I stars, the elements sodium, phosphorous, aluminum, and others, begin to be formed. This process goes on and is responsible for all those elements so essential to our technology.

At the present time our Sun has used up about half of the hydrogen it possessed at its birth about five thousand million years ago. The remainder has been converted into helium. This means that the Sun has still another five thousand million years to go before its demise, which is likely to take the form of transition to a red giant (which will consume Earth and all the inner planets). It will then decline to a hot, very dense, white dwarf. All this is so far ahead in time that it need not cost us a night's sleep! We see, however, how all the essential elements are available for our civilization and for others—for better or for worse!

APPENDIX 2

Throughout the entire length of this book the assumption has been made that life, intelligent life that is, abounds in the galaxy and that we here on Earth are neither alone nor unique in the universe. To those who might find this assertion presumptious we would like to set out briefly and as concisely as possible the reasons so many astronomers now accept the considerable likelihood of alien life forms and highly advanced civilizations.

The necessary prelude was set out in Appendix I, and what follows here is, in fact, merely a logical continuation and, we believe, a reasonably logical conclusion. In Appendix 1 we saw that a planetary system such as our own having an abundance of heavy elements must be formed in association with Population I stars. Since there are around a billion stars similar in type to our Sun within the Milky Way, it follows that there could be around a billion planetary systems approximating our own in the galaxy and anything up to a billion civilizations at varying stages of development, some well behind us, some well in front.

To ascertain why this should be, it is necessary first to study our own planet from a chemical standpoint. It is fairly conclusive by now that in the Solar System Earth is the only planet that supports intelligent life. Indeed it may well be the only planet in the system that supports life of any sort. (At the present stage in terrestrial history there may be those who are dubious about applying the adjective "intelligent" to the life that exists here on Earth—and not without reason.) Was this simply some sort of cosmic accident or was it an essential consequence of the laws of nature once the appropriate physical conditions prevailed?

An initial glance at the position gives the impression that the physical conditions being so demanding and so numerous, the chances of life initiation and subsequent development are exceedingly low and, consequently, the chances of life existing on any other planet within the galaxy remote. This, however, is not so. The relevant physical conditions are not independent of one another. They are interrelated and follow as a natural consequence from the processes that led to the evolution of the planetary system. In other words, physical processes lead logically to chemical processes, and these in turn lead to biochemical processes and ultimately to biological evolution.

A star such as the Sun evolves from a nebula, a cosmic cloud of elemental gases consisting essentially of hydrogen, helium, and the heavy elements, the first two predominating. The gravitational tidal forces that result from the birth of the star cause minor condensations (protoplanets) in the residual surrounding cloud at specific distances from the star. These protoplanets after due time condense into true planets. It may be that Jupiter, Saturn, Uranus, and Neptune have still not done this completely. A zone in which temperature and other conditions are appropriate to life comes into being. This ranges from approximately 70 million miles from the star to about 140 million miles. So far as our Sun is concerned, three planets, Venus, Earth, and Mars lie within this belt.

It is quickly apparent that the fundamental chemistry of those planets within 140 million miles of the star (in our case Mercury, Venus, Earth, and Mars) are totally different from those lying farther out (e.g., Jupiter, Saturn, Uranus, and Neptune). The first four are relatively small solid bodies with clearly defined surfaces. The outer four, on the other hand, are gas giants. The reason for this difference is not hard to find. Hydrogen and helium, which constitute the major portion of the residual cloud, escape easily from the inner planets because their small masses give rise to such low gravitational attraction that they cannot retain the very light atoms of hydrogen and helium. These light atoms are then repelled from the inner planets by emanations from the star.

Inner planets are small, relatively dense spheres with masses and radii roughly equal to those of Earth. They contain very little hydrogen but an abundance of carbon, nitrogen, oxygen, and the heavy elements. So far as life on Earth is concerned this is a highly fortuitous circumstance. Had not most of the hydrogen escaped, life could not have emerged for the simple reason that this hydrogen would have combined with all the free carbon, nitrogen, and oxygen upon which life is based. Moreover, the mass and radius of Earth were just right to prevent the escape of nitrogen, oxygen, carbon dioxide, and water vapor. Without these elements, life would have been impossible. That a planet such as our own is formed at a distance of about 90 million miles from its central star (the Sun) renders it virtually certain that life will emerge, because the chemical and physical criteria for the formation of the molecules upon which life is based are satisfied. Planets like Jupiter and Saturn situated well outside the "life" zone are very massive. They will therefore have retained most of their hydrogen and will, as a consequence, possess little or no potential for life initiation and development. (Notwithstanding this, we have been treated for several decades now to sci-fi Jovians and Saturnians, not to mention Uranians and Neptunians!)

During the past twenty years it has been increasingly evident that life-supporting organic molecules are formed at any point in the galaxy (and presumably in all other galaxies) given the appropriate conditions. In that time molecules of an organic nature have been detected in interstellar space. So far, using radio-telescopes and infrared detectors, astronomers have positively identified more than twenty-six different organic compounds. These include formaldehyde, acetaldehyde, formic acid, and methyl alcohol. Moreover, in the last decade, there have been several instances of meteorites falling to Earth that contained about a score of amino acids—very essential "building blocks" in the production of proteins. There was also clear, chemical evidence that these amino acids were definitely formed in *space* and *not* on Earth after the meteorite's decent.

We see, therefore, in the galaxy and in the universe as a whole, a natural chemical progression from simple atoms to complex organic compounds capable of supporting life under the conditions prevailing on primitive Earth (and on all the other primitive "Earths" in the universe). It is reasonably clear now that terrestrial life was preceded by the formation of organic compounds in the chemically reducing conditions of a primary (hydrogen) atmosphere and that life began when conditions, atmospheric and otherwise, become appropriate to molecular replication. From there on the course for evolving, developing life was clear and open. So it was on Earth—so it must have been on a countless host of other planets in this galaxy and in all the countless others. The universe is not lifeless. It throbs and teems with life at all stages of development—and of civilization! Quite assuredly, we are *not* alone!

BIBLIOGRAPHY

CHAPTER 2

1. Sheafer, Robert. "Project Ozma II." *Spaceflight*. London: December 1975: *17*, 421–423.

2. Sheafer, Robert. "NASA Contemplates Radio Search for Extra-Terrestrial Intelligence." *Spaceflight*. London: October 1976: *18*, 343–347.

3. Ridpath, Ian. "An Ear to the Void." *New Scientist*. London: May 12, 1977: 326–328.

4. Tarter, Jill et al. "Searching for Extra-Terrestrial Intelligence: The Ultimate Exploration." *Mercury*. San Francisco: July/August 1977.

5. Sagan, Carl. "The Search for Extra-Terrestrial Intelligence." *Scientific American*. New York: May 1974: 80–89.

6. U.S. Government Printing Office. "The Search for Extra-Terrestrial Intelligence." Washington D.C.: 1976.

CHAPTER 4

1. Viewing, D.R.J. et al. "Detection of Star-ships." *Journal of the British Interplanetary Society*. London: 1977: *30*, 99–104.

CHAPTER 7

1. Hindely, Keith. "Beware the Big Wave." *New Scientist.* London: February 9, 1978: 346–347.

2. Harvey, J. G. "Waves and Tides." *Atmosphere and Ocean.* London: Artemis Press, 1976: 73–83.

CHAPTER 9

1. Gwynne, P. et al. "Lighting a Sun on Earth." *Newsweek.* New York: November 21, 1977.

CHAPTER 11

1. Valéry, Nicholas. "The Shape of War to Come." *New Scientist.* London: June 17, 1976: 628–629.

2. Hussain, Farooq. "Is Legionnaire's Disease a Russian Plot?" *New Scientist.* London: December 15, 1977: 710.

CHAPTER 13

1. Walgate, Robert. "Russia's Incredible Beam Weapons." *New Scientist.* London: May 19, 1977: 379.

2. Andrews, David. "Picosecond Pulses of the Billion MW Laser." *New Scientist.* London: January 5, 1978: 32–33.

CHAPTER 14

1. *Encyclopaedia Britannica.* "Biological Warfare." 1968: 641–643.

CHAPTER 17

1. Hope, Adrian. "Finding a Home for Stray Fact." *New Scientist.* London: July 14, 1977: 83.

2. Darwin, G.H. "Barisal Guns and Mist Pouffers." *Nature.* London: October 31, 1895: 650.

3. *New Scientist.* "Mystery Booms Haunt American Coast." *New Scientist.* London: February 9, 1978: 341.

4. Ritchie, M. "U.F.O. Sighting in Ayrshire Sends Shivers up the Spine." *Glasgow Herald.* May 2, 1978.

5. Lunan, D.A. "Space Probe from Epsilon Boötis." *Spaceflight.* London: April 1973: *15,* 4, 132.

6. Herbison-Evans, D. "Extraterrestrials on Earth." *Journal of the Royal Astronomical Society.* London: April 1977: *18,* 4, 511–512.

CHAPTER 18

1. Lawton, A.T. "The Nearest Other Solar System." *Spaceflight.* London: April 1970: *12,* 4, 170–173.

2. Ehricke, K. A. "Astrogenic Environments." *Spaceflight.* London: January 1972: *14,* 1, 2–14.

3. Harrington, R. S. and Harrington, B. J.; "Can We Find a Place to Live Near A Multiple Star?" *Mercury.* San Francisco: March/ April 1978: 34–35.

CHAPTER 19

1. Freitas, R. A. J. "Metalaw and Interstellar Relations." *Mercury.* San Francisco: March/April 1977: 15–17.

2. Asimov, Isaac. "The End." *Penthouse.* New York: January 1971.

CHAPTER 20

1. Papagiannus, M. D. "Are We Alone?" *Journal of the Royal Astronomical Society.* London: 1978: *19,* 227–281.

APPENDIX I

1. Motz, Lloyd. "Cosmic Chemical Engineering." *Chemical Engineering.* April 1974: 86–94.

2. Kroto, Dr. H. "Chemistry Between the Stars." *New Scientist.* London: August 10, 1978: 400–403.

APPENDIX II

1. Ridpath, Ian. "Solar Systems and Life." *Spaceflight.* London: August/September 1975: *17,* 323–327.

2. Molton, P. M. "Is Anyone Out There?" *Spaceflight.* London: April 1973: *15,* 246–252.

3. Cottey, Alan. "Advanced Life in the Universe." *New Scientist.* London: April 27, 1978: 236–237.

INDEX